ROCKHOUNDING

— FOR —

BEGINNERS

Your Comprehensive Guide to
Finding and Collecting
Precious Minerals, Gems,
Geodes, & More

LARS W. JOHNSON

with Stephen M. Voynick

Adams Media
New York London Toronto Sydney New Delhi

Aadamsmedia

Adams Media
An Imprint of Simon & Schuster, Inc.
100 Technology Center Drive
Stoughton, Massachusetts 02072

First Adams Media trade paperback edition June 2021

ADAMS MEDIA and colophon are trademarks of Simon & Schuster.

For information about special discounts for bulk purchases, please contact Simon & Schuster Special Sales at 1-866-506-1949 or business@simonandschuster.com.

The Simon & Schuster Speakers Bureau can bring authors to your live event. For more information or to book an event contact the Simon & Schuster Speakers Bureau at 1-866-248-3049 or visit our website at www.simonspeakers.com.

Interior design by Michelle Kelly
Interior images © 123RF, Getty Images, and Wikimedia Commons, and by Frank Rivera

Manufactured in the United States of America

1 2021

Library of Congress Cataloging-in-Publication Data
Names: Johnson, Lars W., author. | Voynick, Stephen M., author.
Title: Rockhounding for beginners / Lars W. Johnson with Stephen M. Voynick.
Description: First Adams Media trade paperback edition. | Stoughton, MA: Adams Media, 2021. | Includes index.
Identifiers: LCCN 2021000311 | ISBN 9781507215272 (pb) | ISBN 9781507215289 (ebook)
Subjects: LCSH: Rocks--Collection and preservation. | Minerals--Collection and preservation.
Classification: LCC QE433.6 .J64 2021 | DDC 552.0075--dc23
LC record available at https://lccn.loc.gov/2021000311

ISBN 978-1-5072-1527-2
ISBN 978-1-5072-1528-9 (ebook)

CONTENTS

Chapter 4: Preparing and Showing Off Your Finds .. 59

PART 2:
ROCKS, GEMS, MINERALS, MINERALOIDS, AND FOSSILS OF NORTH AMERICA / 69

INTRODUCTION

Does the idea of trekking through the mountains to find gold give you a thrill? Maybe you're fascinated by the story held in a fossilized nautilus shell thousands of years old, or eager to share a quartz that you dug out of the ground yourself. Natural treasures are hidden all around, and *Rockhounding for Beginners* is here to help you understand exactly where (and how) to look for them!

Rockhounding is the act of searching for and collecting different rocks and minerals. In the pages that follow, you'll uncover the rich history of this activity and why it's a favorite pastime today. You'll also explore safety tips and the different tools and equipment every smart rockhound has on hand. Then you'll dig deeper into where to look and what exactly to look for, whether you're hunting down gems in a mountain region or fossils along an ocean shoreline. Think of it as your crash course in all things rockhounding.

In Part 2, you'll find profiles for 150 popular rocks, minerals, mineraloids, and fossils. From color patterns and crystal shapes to North American locations, these profiles have all the information a rockhound needs to make the most of their time in the field. Each profile includes a photo of the rock, mineral, mineraloid, or fossil before it's been processed, which will make identifying fresh finds a breeze. You'll also discover helpful tips for collecting, cleaning, prepping, and showing off your prizes.

Maybe it's the idea of searching for buried treasure that first drew you to this hobby. Or you are inspired by the fun challenge of collecting as many different specimens as you can. Whatever your interest, this book has everything you need to find, unearth, and process rocks and minerals like a seasoned rockhound. It's time to seize the adventures waiting just outside your door!

PART 1

How to Rockhound

Where do you look for different rocks and minerals/mineraloids? What supplies should you take with you when rockhounding? How do you clean and store your finds? From hunting for a certain fossil or gem to readying it for display or wearable treasure, there is a lot more to rockhounding than meets the eye. All of these questions may have you feeling unsure of where to start.

Part 1 is here to help! In the chapters that follow, you'll find everything you need to know to get started in the great hobby that is rockhounding. You'll explore how rocks, minerals, and more play into human history and how they have developed for the modern collector. Next, you'll learn about the tools and other equipment every good rockhound has on hand. Once you've got your rockhounding tool kit ready, you'll discover where to look and just what to look for. And locating the different rocks and minerals is just half the fun! After you've uncovered the where of rockhounding, you'll journey through the process of cleaning, prepping, storing, and displaying your finds. Are you ready to begin your quest for the mineralogical treasures hidden right outside your door? Let's dig in.

Chapter 1

THE BASICS OF ROCKHOUNDING

Rockhounding is more than just a fun pastime. It's a practice that goes as far back as recorded history! The earliest civilizations in the world hunted for rocks and minerals, using them to craft everything from weapons to makeup. In this chapter, you'll learn more about the rich past of rockhounding, as well as how that history has shaped modern practices in finding and utilizing earth's hidden treasures. You'll also explore the different reasons why rockhounding is a favorite hobby today—including what may have inspired *you* to learn more. Be warned: Rockhounding can be addictive. You may just want to spend all of your free time out digging for gems.

A Brief History of Rockhounding

People have picked up and coveted rocks and minerals—and the different gems and fossils found within them—since the beginning of time. Who *wouldn't* be intrigued by earth's natural treasures, from sparkling crystals to useful stones to preserved footprints of animals past?

Archaeologists have discovered mines across the globe where early civilizations collected different rocks and minerals. The oldest known mine to date is estimated to be around forty-three thousand years old. It was a hematite deposit used to make ocher, a red pigment used as a paint and makeup. The mines where lapis lazuli and turquoise are collected today from Afghanistan and Iraq, respectively, are approximately five thousand years old. And Neanderthal sites in Belgium have shown evidence of flint mining as early as 4300 B.C.

Early Uses for Rocks and Minerals

What did these people of years past do with their mined treasures, exactly? While there are countless ways in which rocks and minerals are woven into human history, the most common early uses include making tools; weapons; makeup, jewelry, and other aesthetics; and medicine. The following sections explore each of these uses in more detail.

Tools

The oldest known tools date to around 3.3 million years ago and were discovered in Kenya. These tools, made from knapped rock, are more than 700,000 years older than the stone tools found in Ethiopia that were previously believed to be the oldest in human history. It's not yet known what these ancient tools were used for.

From there, the history and scope of man-made tools and the different rocks and minerals they were forged from is vast. Artifacts and records left behind show just how innovative early civilizations across the world were with the buried prizes native to their land. By the end of the Stone Age (twelve thousand years ago), flint was commonly mined throughout Europe for its durability in many tools such as axes, as well as its ability to spark flame. In Scandinavia, Vikings (from the eighth to eleventh century) made an early version of an optical lens out of quartz crystal and are believed to have used what is known as Iceland spar, a variety of optical calcite, as a navigational tool on their boats. And obsidian, an extremely sharp mineral (technically a mineraloid, a mineral-like material), has long been used by many cultures to make tools such as knives and scraping utensils used for cleaning animal hides. It was also once the primary material in scalpels used for eye surgery. Some surgeons still use obsidian scalpels today.

Weapons

The practical uses for rocks and minerals didn't stop with tools. Just as soon as they were developed for harmless uses like cutting and carving wood, they were made into weapons. Sharp obsidian has been popular in blades, arrowheads, and more as far back as the early Stone Age. Flint was also commonly knapped into weapons. And in 2016, researchers discovered that a dagger found in the tomb of the Egyptian King Tutankhamen in 1925 was crafted out of meteorite—mineral fragments from outer space.

Jewelry, Makeup, and Other Ornaments

It didn't take long for people to realize that there was more to rocks and minerals than just materials for practical tools and weapons. Evidence shows that found specimens have been used for aesthetic purposes such as jewelry, makeup, and more since the beginning of human history.

Beads were one of the first artisanal products of rocks and minerals; the practice of bead drilling dates to around one million years ago. Early Hindu and Buddhist rosaries were made with minerals such as agate and cinnabar. Traditional Japanese beads known as *magatamas* were first made with materials such as slate and talc before being designed almost exclusively in jade. Garnet was one of the first faceted beads; Egyptian bead makers were cutting, polishing, and drilling the gem by 3100 B.C. Beads were commonly strung together (or individually as early pendants) to make bracelets and necklaces. Archaeological evidence also shows that Stone Age people wore small fossils as decorative pendants.

Ancient innovators also found ways to decorate their skin directly with different minerals. The famous Egyptian black eye makeup kohl was originally made from metal sulfides such as galena, pyrolusite, magnetite, and stibnite. Many of these minerals contain elements that are toxic to humans, and today, "kohl" is made from safe-to-wear synthetic ingredients. Copper minerals such as malachite and chrysocolla were also used by ancient Egyptians as well as by Sumerians to make green makeup.

ROCKHOUND IN THE KNOW

So what exactly is the difference between a rock, a mineral, and a gem? Often these three terms are used interchangeably in rockhounding, but there are some differences between them:

Rock: A rock is a mineral aggregate (combination of minerals) that doesn't necessarily have a specific chemical composition. There are three types of rock: sedimentary, igneous, and metamorphic.

Mineral: A mineral is a naturally occurring, solid, inorganic, crystalline material with a definite chemical composition. While rocks have three main types, minerals come in thousands of forms.

Gem: A gemstone is any rock or mineral that has both aesthetic and economic value.

Besides their creative makeup, ancient Egyptians were known for using minerals in many ornaments and decorations. For example, the tomb of King Tutankhamen was found filled with not only gold but many gemstone artifacts. The boy king's solid gold burial mask included inlays of lapis lazuli, obsidian, quartz, carnelian, amazonite, and turquoise. A breastplate was also discovered in the tomb bearing a scarab carving made from a variety of tektite known as Libyan desert glass.

Medicine

Ancient Egyptians also used minerals for medicinal purposes. Malachite and chrysocolla were used to cure ailments such as abdominal or dental issues. The copper powder made from these minerals was also used to clean wounds. Copper has since been proven to have antimicrobial applications. Sometimes gems and minerals were worn as talismans by many peoples worldwide to ward off both physical and spiritual ailments. The early Chinese wore polished jade amulets to assure health and longevity. And in the Baltic region, amber amulets were thought to protect against diseases, while powdered amber was applied to promote healing of wounds and infections. In medieval times in western Europe, sapphire amulets were worn, or powdered sapphire was ingested, to alleviate the discomfort of rheumatism. In early China and medieval Europe, the ingestion of powdered fossils was thought to assure longevity.

Storytelling and Research

Beyond mere ornaments or offering the possibility of a longer life, fossils provide a glimpse into the past that was not lost on early civilizations. Some scholars suggest that the monsters of Greek mythology were inspired by the fossil remains of large prehistoric creatures. And in the early 1800s, curiosity about the origin of fossils inspired the beginning of paleontology—the study of life-forms in bygone geologic eras. Just as amazing to these early rockhounds, collectors, and accidental discoverers as a shiny gem was the preserved footprint of a creature that once roamed the earth.

Rockhounding in North America

When European settlers arrived in North America, they wasted no time finding valuable rocks and minerals to continue in their countless practical and artisanal pursuits. In fact, in the 1960s it was discovered that Viking settlements in Newfoundland (dating back to A.D. 1000) were built near peat bogs containing bog iron, a type of iron ore that forms when rivers carry dissolved particles of iron into wetlands. The Vikings used bog iron to make nails for their ships.

In the 1400s, Italian explorer Christopher Columbus promised the Spanish Crown gold. Of course, while he intended to reach India, his discovery of the islands off North America eventually brought many more Europeans looking for precious metals. Settling in the Eastern US, these early colonists mined for metals that were sent back to England.

Over time, more and more of Canada and the US were colonized, and with this growth came more mining—and more mineralogical discoveries. In 1847, gold was discovered in California, and in 1859, it was discovered in British Columbia, sparking gold rushes that fueled much of the expansion across North America. Many seeking their fortune discovered a continent full of natural wonders along with gold. The rocks, minerals, and fossils found as North America was settled were endless. Everything from useful marble to precious emeralds to proof of dinosaurs was uncovered in the quest for a prosperous life in the New World.

As time went on, these treasures became more than just a way to make a living. By the 1930s, rock and gem clubs started forming around the United States, bringing together people interested in collecting rocks and minerals for many different reasons. Some of these clubs still meet to this day!

New minerals and fossils are still being discovered in North America. In 2018, the most complete skeleton to date of an Enantiornithine bird, one of North America's largest dinosaur-era birds, was uncovered in the Grand Staircase-Escalante National Monument in Utah. And as recently as 2019, a new sapphire deposit was discovered in British Columbia. Only time will tell what new treasures await you out in the field.

Why Rockhounding?

People get into the hobby of rockhounding for just about as many reasons as there are rocks and minerals. Rockhounding is a mostly outdoor activity that is a lot of wholesome (sometimes dirty) fun. If you enjoy being out in nature, rockhounding is likely to be right up your alley.

Some people will plan a whole trip around digging at one or several rockhounding locations. Sometimes it's added as a detour or complement to a more traditional outdoor activity such as fishing or hunting. Hunters often accidentally discover mineral deposits because they tend to go off trail while tracking game. Others may spend free time looking for rocks while on a destination vacation or during business travels.

Other Interests

Rockhounding can also lead people to other outdoor activities. Some rockhounds will plan their searches in locations that are near hot springs, waterfalls, or other natural wonders, so they can enjoy a relaxing soak, refreshing swim, or other fun experience after digging for specimens.

FROM ONE ROCKHOUND TO ANOTHER

I once visited a thunder egg deposit that was just okay. It was nice to check the site off my list, but the thunder eggs I found were nothing special. However, while digging at the deposit, I noticed that morel mushrooms were growing everywhere. I quickly gave up digging for thunder eggs and picked several pounds of mushrooms that I shared with friends and family later that evening. Now I visit that site mostly to pick mushrooms.

You also can't help but notice the different animals you see while out in the field, and this has inspired many a rockhound to learn more about animals in their free time. The same things can be said about plants, geology, weather systems, astronomy, petroglyphs, local history, and so much more.

Growing Collections

Rockhounding appeals to collector types. Some people want to get one of every rock, mineral, or gem they can get their hands on. Other collectors specialize in certain minerals such as agate, quartz, or fluorite that can form in a wide array of crystal habits and colors. This type of collecting can also be specific to locations. For example, one fluorite collector might focus on obtaining a specimen from every known fluorite location in North America or even the world. Another fluorite collector could spend a lifetime collecting the various formations and colors that come from one specific mine, region, or country.

The Rockhounding Rush

The thrill of the hunt is what drives some. At first, just finding a site listed in a guidebook is exciting. When you spend time scoping out an area, just knowing you've made it to the theoretical (or maybe literal!) *X* on the map can be exhilarating. Then you get to hunt. In some areas, finding one or two good pieces may be a great day. At other sites, you'll walk away with full buckets of specimens. Either way, uncovering mineralogical treasures is quite the rush. It feels amazing to unearth something no one else has found.

Final Notes

Rockhounding is a great lesson in patience and enjoying yourself, regardless of whether or not you find the rocks and minerals you are looking for. The patience comes from time spent prospecting and collecting, and the fact that sometimes you might put in a lot of effort for little return.

When you don't find any specimens, it's called getting skunked. One of the most important parts of rockhounding is accepting that you're likely to get skunked every now and then. For that reason, it's important to appreciate the process, as well as the little opportunities rockhounding has to offer. Just the fresh air and natural sights all around are reasons to feel satisfied with an experience. Don't give up if you aren't finding the same quality or amount of material you may see others sharing in clubs or online groups. Like with many other pursuits, you'll only get better at it with experience. Try to avoid comparing your experiences with those of other rockhounds.

What Do You Want to Collect?

What are *you* looking to find out there? Dozens of rock types and over five thousand minerals have been found on Earth—and many have different varieties. Thousands of fossils have also been discovered, and new fossil species are being found all the time. In short, you can collect a lot of different rocks, minerals and fossils!

When first getting into rockhounding, it's a good idea to spend time finding a variety of different rocks and minerals. You may discover you love finding them all. You may realize that agates are the most fascinating to you, or that uncovering fossils is your true passion. Dabble a bit in every area to find what suits your interests best.

When you determine what you really like, you can start getting into specific collections. You may spend all of your time collecting just one mineral in its various crystal habits, or one mineral from different sources. Some people find joy in collecting large specimens, while others collect tiny specimens called micromounts. There are people who don't care what mineral they find as long as it's heart-shaped. Many only collect material that they can use for lapidary work such as making cabochons or spheres, or faceting. There is no limit to the types of rock and mineral collections you can create.

Safety First

Now that you've explored the history and reasons for rockhounding, you're ready to learn about the tools and equipment you'll need—*almost*. Rockhounding is a fun and highly rewarding pastime. That being said, you will want to keep in mind a few key factors as well as precautions to ensure a positive, safe experience.

Weather

Know what weather conditions to expect so you can pack and prepare accordingly. Snow can block the road to a site, so it's handy to know the snow level at a given location. Heat can make it unsafe to dig for long periods of time, so be prepared to provide yourself some shade, have plenty of water on hand, and build in breaks between digging. I've seen people go out to the high desert in the summertime wearing only shorts, thinking a desert is naturally hot, only to find out that the desert can get very cold at night. Rain in certain areas can make it dangerous to drive to a site; you may run the risk of getting stuck in mud or sliding off the road along the way. It's always best to play it safe.

Animals

North America has a wide variety of animals, and spotting them in the wild is one of the many bonuses to rockhounding. However, some can be dangerous. Even if you're unlikely to see a dangerous animal, let alone have any sort of problem with one, it helps to be prepared. Be sure to read the following details on specific North American animals and what to do should you encounter them.

Mammals

North America is host to many large predatory mammals, including bears, mountain lions, bobcats, wolves, and coyotes. While these animals tend to stay away from humans, you can never be too careful or prepared. Use proper food-storage techniques, especially in an area known for bear encounters. The scent and accessibility of food can attract these curious animals.

It is always wise to do some research about what kinds of wild animals you may run into in the area you are planning to visit. Look online for common predators and what measures you can take both before your trip and in the event of an encounter.

During your rockhounding adventure, always be aware of your surroundings. It can be easy to get so caught up in what's on the ground in front of you that you don't notice a large animal nearby.

Most animals will want nothing to do with you but can get very aggressive if they are injured or sick, or if you get too close to their babies. Always stay calm and try to slowly remove yourself from the situation. You may want to consider having a defensive weapon on hand such as bear spray.

Large nonpredatory mammals such as moose, elk, and deer can also be dangerous. It is best to give these animals distance. They are likely more scared of you than you are of them, but they can hurt you if frightened or protecting young. Stick to taking photos from afar.

And while less of a physical threat, vermin such as rats and squirrels can carry diseases such as hantaviruses. Avoid touching rocks or other specimens that have excrement on them, and do not approach any vermin you may encounter.

Reptiles

Venomous snakes can be found all over the continent. The most feared is the cottonmouth, but rattlesnakes and the eastern coral snake are dangerous as well. Keep in mind that many snakes that are not venomous may still lash out if they feel threatened. While rattlesnakes tend to give themselves away with their distinct tails, it helps to know how to identify the different venomous and nonvenomous snakes and how to treat a snakebite by either type. Pack any recommended first aid in case of a bite. Gila monsters and Mexican beaded lizards can be found in the southwestern United States and are also venomous. Look online for more information on what to do in the event of a bite.

Insects

There are plenty of creepy-crawly insects in every corner of North America that can be poisonous, painful, or just downright obnoxious to be around—as many a rockhound can attest.

FROM ONE ROCKHOUND TO ANOTHER

The area around the sunstone mines in Oregon is well known for its fluorescent scorpions. Most of my many trips to the area have been in May or June, when it seems impossible to find even one scorpion. However, one trip I took to the area in September yielded different results. At night, I took a black light around the base of the sagebrush where I had set up camp and found that every single bush had anywhere from four to seven small scorpions under it. Keep in mind these scorpions' stings aren't much worse than a beesting, but I still checked my boots thoroughly in the morning.

Many insects only come out at certain times of year, so if you have any phobias or real concerns about them, be sure to schedule your rockhounding adventures accordingly. Regardless, it is always a good idea to bring along bug deterrent, first aid supplies, and any allergy medications needed in the event of a bite or sting. Know whom to call in the event of an emergency such as a venomous spider bite or surprise allergic reaction.

Plants

Many plants across North America can be problematic to the rockhound. Plants such as poison oak, poison ivy, wood nettle, stinging nettle, poison sumac, and ragweed cause minor to severe skin irritations when touched. Brambles, devil's club, and cacti can all be unpleasant as well. Wear proper clothing if you expect to be poking around near potentially harmful plants. Many plants are also poisonous; never eat anything you find in the wild unless you are one hundred percent sure it is safe.

Tools

There is a proper way to use each implement you bring along to aid in your rockhounding adventures. Know your equipment and how it's meant to be used to avoid any potential malfunctions or injury. For example, never use a regular carpenter's hammer to smash rocks. They are not tempered the same as geology hammers and can splinter, sending shards of metal everywhere. More rockhounding tool and equipment specifics are covered in the next chapter.

People

Whether rockhounding close to home or in a more remote location, be aware of your surroundings at all times. You're not always alone. You may accidentally stumble onto private land or a digging site someone else has claimed. It's your responsibility to know where you're standing and to ensure those around you are not a potential danger. Depending on where you choose to dig, you may want to consider bringing along a rockhounding buddy or two as well.

Be Prepared for Anything

Murphy's Law applies to rockhounding as much as it does to any other aspect of life. Rockhounds share dozens of stories about unexpected things that have happened while out in the field. While many make for a good laugh, others are less pleasant to remember.

At the end of the day, preparation will be your best friend in having a fun, safe rockhounding experience.

FROM ONE ROCKHOUND TO ANOTHER

I was once leading a short rafting trip with a group of rockhounds looking for agate, jasper, and petrified wood. Toward the end of the journey, a friend got out of his raft to pull it to shore. The water was murky, and visibility was none. He ended up cutting his leg on a piece of sharp metal fence post that had made its way into the water. Luckily, we were prepared at camp with plenty of first aid supplies to stop the bleeding before the long drive to the nearest hospital.

Rules and Regulations

The rules and regulations for where you can travel and what you can do in certain areas differ from state to state and from province to province. It's your responsibility to know what land you are on and the laws pertaining to that land.

Obviously, always stay off private property unless the landowner gives permission. Fortunately, most accessible rockhounding sites are found on public land. In the United States, these lands are run by the county, state, or federal government. Not all public land in the US is available for rockhounding, and collecting is not allowed in some state and county parks, wildlife refuges, national parks, national monuments, tribal lands (if not a member of the tribe), and wilderness lands. Look online or call ahead to ensure you are allowed to rockhound in a chosen location.

In Canada, public land is referred to as Crown land and is administered by the provincial government. The rules for rockhounding on Crown land are similar to those on much of the federal land in the United States. Minerals, rocks, and certain fossils can be collected for noncommercial, personal, or recreational purposes. Only manual collecting methods may be employed, and strict quantity limits are imposed. These rules are enforced by individual Canadian provinces. Check online for more details about the restrictions in a certain area before collecting.

The rules about what you can dig up and how much you can take also vary from place to place. More often than not there are regulations on what fossils you can collect. Some areas also have weight and size limitations on what rocks, gems, minerals/mineraloids, and fossils you are allowed to collect and how much you can collect per day and per year. Depending on where you are and what you're collecting, a permit may be required to collect rocks, gems, minerals/mineraloids, or fossils. Always check the collecting laws for where you plan to rockhound.

Next Steps

Now that you've journeyed through the history of rockhounding and what it has to offer, and also taken a closer look at the important precautions for a positive *and* safe experience, you're ready to assemble your rockhounding toolbox. In the next chapter, you'll learn about the different tools and equipment needed to find, dig up, and separate rocks and minerals in the field. Remember: Preparation is key. So read on, rockhound!

Chapter 2

ROCKHOUNDING EQUIPMENT AND TOOLS

As a rockhound, it's important to be well equipped for anything. You never know what Mother Nature will throw at you! You also never know what you may find while in the field, and having the right tool to extract your discoveries is crucial. In this chapter, you'll explore all of the main equipment and tools you will want to have on hand during your rockhounding adventures. From recommended safety gear such as eye protection and gloves, to collecting tools such as pry bars and brushes, you'll never have to leave a treasure behind because you weren't ready for the job. You'll also discover valuable resources such as field guides, ultraviolet lights, and record-keeping aids that will help you navigate to a site, identify specimens, and keep track of your finds.

Attire and Safety Gear

The proper clothing and safety gear are crucial elements for a productive, enjoyable rockhounding trip. The right outfit can make or break your day of rockhounding.

FROM ONE ROCKHOUND TO ANOTHER

Make sure to put on a proper pair of kicks before heading off into a desert, forest, mountain, or other unpredictable terrain. It is all too tempting to focus on comfort and neglect practicality, especially if your chosen rockhounding site is far away, but don't fall for this trap. I have twisted an ankle on more than one rockhounding adventure by opting for casual sneakers with no grip over a sturdy pair of hiking boots.

The following sections explore proper rockhounding attire and safety gear in more detail.

Clothes

When you're out in nature, the weather can change at a moment's notice. For example, the desert can be blazing hot during the day and almost freezing as soon as the sun goes down. Sudden desert storms can catch a rockhound off guard. So be prepared for it all. Check the weather predictions before heading out, but also pack or wear the following:

- A raincoat
- A durable pair of pants to protect your legs when digging and hard-rock mining
- A hat to shield you from the sun
- A warm sweatshirt or jacket in case of a temperature drop
- A long-sleeved shirt and long pants for potential sandstorms

Wear what is comfortable for you, but always take into consideration what sorts of specimens you will be looking for and where you will be doing your rockhounding.

Footwear

Like clothing, the proper footwear depends on what you plan on doing and where. If your rockhounding trip requires you to do some hiking, a sturdy pair of hiking boots is essential.

When exploring watery areas like rivers, lakes, or oceans (or rainy days on land), you'll want to sport some waterproof shoes, especially if the water is cold. Rubber boots or waders are popular options if you expect to get your feet wet or need to cross shallow creeks or rivers. In warmer waters, a pair of durable sneakers you don't mind getting wet or some aqua socks are perfectly acceptable.

In general, you should avoid wearing open-toed shoes when rockhounding. If you do choose this option, however, be extra careful to watch where you step, and do not attempt hard-rock mining.

Eye Protection

If your rockhounding trip involves pounding on rocks with a hammer, proper eye protection is essential. An eye injury is one of the last things you ever want happening to you—especially when out in the field without quick access to a healthcare facility. Find the protective goggles or glasses that feel most comfortable and give you complete eye coverage. Regular prescription glasses or sunglasses are better than nothing but are not enough to protect eyes from pieces of flying rock.

Gloves

A thick pair of gloves can be not only downright handy but sometimes necessary when rockhounding. Obsidian shards can be extremely sharp. Many other minerals may have very sharp edges as well and can cut or stab your skin easily. Gloves also help protect your fingers when moving rocks on overburden or talus hillsides. It's also a good idea to wear gloves while hammering rocks to cut down on the numb sensation that can come with extended periods of hammering.

First Aid Kit

A proper first aid kit is the most important thing to have with you when rockhounding. It can truly make the difference in an emergency situation. The most popular item in a rockhound's first aid kit are bandages. Scrapes can go hand in hand with digging and hammering. Your kit should also include:

- Gauze
- Wraps
- Trauma pads
- EMT shears
- A splint
- Alcohol swabs
- Antiseptic wipes
- Antibiotic ointment
- Safety pins
- Tweezers
- Cotton swabs
- Medical tape
- Nitrile gloves
- Any necessary allergy medication
- Medications for itches, pain, fever, inflammation, and diarrhea

A snakebite kit is advisable when rockhounding in areas with venomous snakes.

If you plan on doing some serious hiking while you rockhound, not only is having a first aid kit essential, but a survival kit is also a good idea. Even the most experienced outdoor enthusiasts can get lost or hurt. It pays to be prepared.

Bug Spray

Bugs are almost inevitable when rockhounding in the field. Some people swear by products containing DEET. Others prefer homemade repellent. Find what works for you and don't leave it at home. Being attacked by tiny things constantly trying to bite or sting you can ruin a perfectly good rockhounding trip.

Sun Protection

Most rockhounding is done in fair to hot weather. Sunscreen with an SPF of 30 or higher is a must for long exposure to the sun, but also on cool and cloudy days. Rockhounds who plan on digging in the same place for a long time often erect pop-up tents or string-up tarps as shelter from UV rays. And don't forget sunglasses. Not only are they helpful in blocking the sun's rays, but some polarized varieties can improve your vision when hunting for specimens in rivers, creeks, and lakes.

Navigation and Record-Keeping Tools and Equipment

A lot of work often goes into planning a successful rockhounding trip. From finding the best route to a site to being ready for any unexpected detours along the way, a little preparation and a few helpful navigational tools can ensure you get to your destination. It's also a great idea to keep records along the way so you can remember any issues you might have had getting to a location, as well as what exactly you found there. The following sections explore the best navigation and record-keeping tools and equipment in more detail.

Notebook and Pen/Pencil

A notebook and writing utensil are handy to have both when planning a rockhounding trip at home and when in the field collecting. Although you can use your phone to make lists too, experience has shown that it's not always easy to use a phone in the field. Your hands may be too dirty to operate the touchscreen and buttons, and bright sunlight can make the screen difficult if not impossible to see.

You may run into other rockhounds in the field who have valuable information to a new or better site nearby. Write down the directions! Even if you think you'll remember, three turns into the adventure, you may just forget if the fourth turn is a right or a left.

Compass/Altimeter

Having a compass while out in nature is always a smart idea. It can be easy to get disoriented while collecting minerals in the field, especially when surface collecting at a vast site where you wander a lot and are mostly staring at the ground. Always know the general direction (north, south, east, or west) you walk from your vehicle toward a rockhounding location.

Altimeters are great to have when your rockhounding trip takes you up a steep hillside or mountain and you need to find a deposit at a certain altitude. An altimeter saves you a lot of hassle of potentially walking too far or not far enough.

GPS

Even with the best maps and directions, it can sometimes be difficult to pinpoint a small deposit out in nature. A GPS can save a lot of time. Luckily, phones and many vehicles come with a built-in GPS. Just enter a waypoint and you're on your way. A handheld GPS is recommended if you are going to walk a fair distance from your vehicle.

Many guidebooks list GPS coordinates for sites, but not all books use the same GPS format. Make sure to match your GPS unit's format with the one used in the book you choose to follow.

Maps and Guidebooks

No good rockhound is without a plethora of maps for both roads and geology. In fact, how to read a geological map is one of the most important rockhounding skills. Knowing what minerals can form where is a big step ahead when rockhounding.

Be sure to bring traditional maps with you as well, as phones and GPS units can run out of battery life. Sometimes the road you need to find isn't on a digital map. Map apps on your phone will also stop working properly when you lose the cellular signal.

Guidebooks are also very helpful, especially when you first get into rockhounding. It's not easy to know where to go when starting out. Guidebooks provide all the information you need, including maps, GPS coordinates, and other details for a successful treasure-hunting outing. Many guidebooks are written about rockhounding sites found in specific states and provinces.

Camera

Rockhounding can take you into some of the most beautiful locations you will ever see—and you're going to want to remember them. From stunning landscapes to flora and fauna, there's so much more to rockhounding than just the rocks. Be sure to bring a camera to capture the wonderful places and sites you'll see. Luckily, most of us have a camera built into our phones, which makes it so much easier to record those special moments. If you do plan to bring a traditional film or digital camera into the field, take into consideration the elements of where you will be rockhounding. A camera can get ruined in a site where it is exposed to too much rain or dropped in water. Sand can also get trapped inside.

Identification Tools and Equipment

With time and experience, you will be able to identify many rocks and minerals just by sight. However, even the most experienced rockhound can get stumped, and a well-stocked arsenal of books and identification tools is helpful whether you're new to rockhounding or years into the hobby. Read more about the best identification tools and equipment in the following section.

Field Guides and Identification Books

There are almost as many rock and mineral guides as there are rocks and minerals. Some guides cover rocks and/or minerals worldwide while others focus on a particular state or province or even a single site. Most guides include the same basic scientific facts, but every guide has some tidbit of information the others don't. When looking for a guide, pick one with good color photos to aid in identification. Keep in mind that it's almost impossible to have a photo of every representation of each mineral in a single publication. A book featuring every formation of quartz would be huge. Start with a good all-around guide and add more specific books over time.

Loupes and Magnifying Lenses

Sometimes magnification is required to view the smaller details of a mineral that help with identification. They allow you to see trace minerals, grain surfaces, small crystals, details in crystal structure, and microfractures.

Loupes (also known as hand lenses) are small magnification tools that are easily carried in your pocket. They come in 10x, 15x, or 20x magnifications. The weakest magnification, 10x, gives you the best depth of field and light capture, while 20x is better for viewing very fine details.

Ultraviolet Lights

Some rocks, gems, minerals/mineraloids, and fossils will fluoresce when exposed to shortwave UV light, long-wave UV, or both. Some rockhounds focus their entire collection around fluorescent minerals. Serious collectors often build their own UV boxes or even entire rooms to display their glowing rocks.

ROCKHOUND IN THE KNOW

There is a type of sodalite-rich variety of syenite, an igneous intrusive rock, found on the shores of Lake Superior that when exposed to long-wave UV light glows a bright yellow that makes the stone look almost as though it's on fire.

Many specimens that fluoresce often don't look like much in the regular-light spectrum. For this reason, they are collected at night with a handheld UV light.

Collecting Tools and Equipment

Of course, you can't properly unearth and collect the specimens you identify without a few essentials. The following sections explore the recommended collecting tools and equipment for rockhounding in more detail.

Geology Picks

If there is one quintessential tool synonymous with rockhounding, it's the geology pick (also known as a geologist's hammer, rock hammer, or rock pick). A silhouette of a geology pick is used as a logo by almost every rock club and many rock shops to a point that it's almost cliché. Rock hammers usually have two heads, one on either side. One head is flat, square, and used to break rocks and to hammer in chisels (also called gads). The other side is either a chisel or pick-shaped. The chisel head is used for clearing loose rocks, dirt, and vegetation. It is also useful for splitting slate and shale when looking for fossils. The pick head is used for prying and digging, as well as for splitting rocks. Most geology picks are about 13" long.

Geology picks differ from regular hammers in that the steel is specially tempered to pound on rock. Geology picks are also made from a solid piece of steel, versus a steel head attached to a handle of a different material.

CAUTION

Regular hammers can splinter, throwing shards of shrapnel everywhere if used on rocks. The hammerhead is also generally attached to a wooden or fiberglass handle, which can break when used on rocks, sending the head of the hammer flying. Never use a regular hammer for rockhounding.

Spend the money for a well-made tool. With the proper care you can have that geology pick for life. Coat your tool with a water-resistant oil to prevent rusting, and sharpen it regularly.

Crack Hammers

Geology picks are perfect for breaking rock, but if you need to smash up large or especially hard rocks, you'll want to call in a crack hammer, a heavy hammer made for breaking rocks and for chisel work. Most rockhounds choose to use what are known as hand sledges. These are shorter-handled crack hammers that come in various sizes and weights. Most are 2–4 pounds and 10"–12" long.

Chisels

Sometimes referred to as gads, chisels are pounded into rock with a rock hammer to break the rock apart. They can be used to trim specimens or to break open larger rocks or geodes. Chisels come in various sizes and tip shapes. Chisels with a flat, straight edge are used to break rocks along a line. These chisels are often used for geodes and to trim specimens. They can also be used to help expose natural fractures in large rocks. Chisels with a pointed tip concentrate the force of the hammer on one focused spot.

There are two schools of thought when it comes to buying chisels. You can buy cheaper chisels because you expect to lose or break a lot of them, or you can spend more money on higher-quality chisels that will last for a much longer time (but sting a bit more when lost). The choice is yours.

Pry Bars

Pry bars and crowbars are used to dislodge and move large rocks. They range in length from 18" to several feet.

ROCKHOUND IN THE KNOW

A great tool for the rockhound is the Estwing Gad Pry Bar. It is an 18"-long piece of steel with a pointed tip at one end and a flat chisel head at the other. It works well with hard-rock mining and is especially handy as it is multiple tools in one. If I'm predominately surface collecting in dirt or gravel, I will carry an Estwing Gad Pry Bar instead of a geology pick, as it's a bit longer than a pick and avoids back strain when I bend over to poke at a rock.

Having a few sizes of pry bars available to fit each job is recommended.

Shovels

Shovels are used to dig holes at rockhounding sites. They come in both long- and short-handled varieties. The long-handled shovels are for digging bigger holes in more open areas. The short-handled shovels are great for tight spaces and when you don't need to dig a large hole. Most shovels used for rockhounding have a spade head, but flat-head shovels can be useful for scraping up the last bits of pay dirt, or gem-filled concentrate, off the ground. When it comes to shovel handles, fiberglass is a stronger option than wood.

Screens

Sifting screens are used to separate dirt from rocks and for classifying sizes of rocks. They are made from galvanized hardware cloth. The most common sizes of hardware cloth have ¼" and ½" mesh openings, but larger and smaller screens can be used depending on what you are looking for. I have a small 12" square screen with a ½" mesh that I use for small garnets.

The frames of most screens are made of wood, but some screen frames are made from PVC piping or aluminum. Smaller screens can be easily used on the ground. For larger screens, you may want to look into buying or building a stand to rest the screen on so you don't have to bend over to shake the screen. Tripods that hang a screen also work well.

Screens made of plastic that are stackable are used for gold panning. They allow you to classify and separate larger rocks and nuggets from fine gold.

Screens can easily be made at home or purchased online. Some rock shops sell them as well.

Gem Scoops

A gem scoop is a tool made of a long handle with a slotted scoop at the end. It is used to scoop up mineral specimens from gravels and dirt. Gem scoops are most useful when exploring gravel deposits along oceans, lakes, creeks, and rivers. Gem scoops are lifesavers when it comes to avoiding back strain. Long hours of bending over to pick up rocks wreak havoc on the lower spine, and gem scoops allow you to stand up straight or bend just slightly.

Most commercial gem scoops are made from aluminum, so they aren't heavy and even make for great walking sticks. The scoop isn't welded to the handle, so don't use one to pry heavy rocks. If you notice your scoop has become a bit loose over a long period of use, you can pinch the metal again where the scoop is attached to the handle.

Probes

A probe is a long piece of metal bent at one end into a handle. Agate and jasper hunters use them to poke in the dirt to find the minerals. This is because agate and jasper make a particular "ping" sound when struck by metal. If you hear this sound, it's a sign of a good place to dig.

Buckets

Five-gallon plastic buckets are one of the most commonly used tools to collect and store rocks in. Buckets are convenient because they have handles and are easy to obtain, and a full 5-gallon bucket is about the limit of what most rockhounds can carry. If you have trouble picking up a full 5-gallon bucket, try using a 1- or 2-gallon bucket.

For easier bucket transportation on flat surfaces, try using a Broll. It's a contraption with wheels and a long handle that you attach to a 5-gallon bucket. Rolling a bucket full of rocks is much less strenuous than carrying the bucket.

If you plan on using buckets for outdoor storage, it's a good idea to drill some small holes in the bottom of the buckets to allow water to escape. A forgotten bucket of rocks can get coated in slime and algae if unable to drain during rainfall.

Packs and Bags

Heavy-duty backpacks and bags are another common item used to move rocks around in the field. Make sure that the bag you choose is made of strong material that can withstand the weight and pointy bits of rocks without tearing. A backpack with a metal frame helps distribute the weight of rocks on the back.

A cheaper option for lugging rocks around is sandbags. You can get them in bundles of one hundred, they're very strong, and when empty they take up much less room than buckets.

Spray Bottles

Spray bottles are used to clean dirt off specimens in the field. When rocks are wet, you can get a better idea of what they will look like when polished. Spray bottles are good for people too. A nice mist of water on your face in a hot desert can be one of the best feelings you have ever known. (Make sure you also have plenty of water for drinking!)

Brushes

Brushes are another item that help clean dirt off specimens when you're in the field. They come in various sizes and shapes with a variety of handles. The bristles can be made of plastic, steel, aluminum, or brass. Metal bristles can damage some softer minerals, so if you're not sure about the hardness of the mineral you're cleaning, use a scrubbing stick with a plastic bristle brush. Pro tip: Save old household cleaning brushes and toothbrushes to repurpose for scrubbing rocks. Paintbrushes are useful for sweeping away dirt in a dig hole or vug.

Gold/Silver/Platinum Tools

Many books have been written over the years about the tools used for precious-metal mining. For the beginner gold hunter, a gold pan, sluice box, or metal detector are basic tools to start with. If you catch the "gold bug," you can research more advanced tools and machinery.

Gold pans are the oldest method of recovering gold from waterways. Gold is very heavy, and by shaking and swirling material taken out of a river or creek, any gold will ideally work its way to the bottom while lighter rocks are floated out of the pan. When you are done, you should be left with heavy materials such as black magnetite sand, garnets, and hopefully precious metals. You will then need to separate the gold from the rest of the heavy material.

Another method for separating gold from dirt is using a sluice box, which is a long, narrow box with bumps or riffles along the bottom. The box is placed in a river or creek, and water is used to pass gold-bearing material through one end and push all the lighter material out the other.

Metal detectors can be used to locate larger gold nuggets and are especially handy when hunting for gold in arid regions where washing dirt with water is not an option.

Readying Your Toolbox for the Field

With a better understanding of rockhounding in your back pocket and the tools for the job in your own rockhound's toolbox, you are well on your way to finding buried treasure! Up next, you'll discover exactly where to look and what to look for, whether rockhounding in a forest, on a mountain, or along a river or ocean.

One important note before you move on: Along with the information in this chapter, be sure to read any included instructions and cautions that come with the equipment and tools you plan to use. Remember that safety is key in any positive rockhounding adventure.

Chapter 3

WHERE TO LOOK IN THE FIELD

Now you know why rockhounding is such a favorite hobby. And you have all the tools and equipment for your own rockhounding adventures ready to go. But where to look? What to look for along that river? What specialized tools might you need for a mountain dig? This chapter is here to answer all of these questions and more. In the pages that follow, you'll explore all the important information and bonus insights for each main rockhounding location—from digging in a forest to panning along a creek. This chapter has everything you need to find specimens like an experienced rockhound.

Tools and Tips for Every Setting

There are many ways one can go about collecting rocks and minerals. It can be as easy as walking around and picking them up or as complicated as setting up a claim and using equipment to dig and extract specimens. The average rockhound will stick mostly to surface finds and digging small holes with simple hand tools. The following sections outline the typical process of rockhounding in creeks and rivers, oceans and lakes, mountains and forests, and deserts in more detail, including using maps to find a site and the specific tools and best times to rockhound in each setting. Be sure to check out the Appendix at the back of the book for recommended online resources and apps to aid in your adventures.

Rockhounding in Creeks and Rivers

Round-tumbled stones have been collected from rivers for thousands of years. The rivers tumble hard minerals and deposit them far away from their source—sometimes hundreds of miles. Many varieties of agate, jasper, flint, chert, geodes, sapphires, topaz, gold, fossils, and much more can be found when hunting along rivers. Bends in the river create gravel deposits during higher water levels. These are the places you'll want to start.

Using Maps

Once you've decided on a creek or river you want to explore, you'll need to find a road map. You can use the old-school road atlas typically stored somewhere in a vehicle, but there are also many options online. And online satellite mapping makes it much easier to determine where gravel deposits will be.

The gravel isn't always easily accessible, however. You may drive miles to check out a gravel bar only to find that the road is several dozen—or several hundred—feet above the river or creek. Luckily, rivers are long, and there are always more spots to check out. On the other hand, large gravel bars that aren't so close to the road sometimes have well-maintained paths leading to them. These paths are not always marked or near a pullout on the side of the road where you can park, so it may take some exploring during your first trip to find the best way to access a deposit.

Finding the Best Place to Stop

One challenge with creek and river collecting is that the best gravel bars always seem to be on the opposite side of the waterway. Late in the summer season, the water is sometimes low enough to easily wade across (especially if exploring a shallow creek), but most of the time it's dangerous to cross.

When exploring a creek or river that you're unfamiliar with, it's always a great idea to bring a friend who has a good eye and can pay attention to the waterway while you drive. (The roads along many rivers in particular can be quite windy, and the driver's eye should always be on the road itself.) The navigator is there to watch the creek or river and look out for any potential gravel deposits, pullouts, and trails.

What to Look For

Once you get to a gravel deposit, it's time to explore. Different minerals are collected differently. Agate, jasper, petrified wood, flint, chert, and fossils are going to be found as float in the gravel deposit. Most silicate materials such as agate and petrified wood will be shinier than most of the other rocks found in the creek or river and are easily identified. Agates will also glow nicely in the sun or when wet. Rainy days can make for good agate hunting on rivers. You can also sometimes splash a bit of water from the waterway onto the gravel to find material you didn't see when the gravel was dry.

Heavier minerals and metals such as sapphires, garnets, fossils, and gold tend to sink into the creek or river and will need to be screened or panned out. Look for bends where the water slows down and heavy material is dumped out. These types of minerals will also concentrate behind large boulders. In the southeastern United States, many people snorkel and even scuba dive in creeks and rivers to find fossils. Gold hunters dive in rivers for material as well.

Many people will also collect river rock and columnar basalt to tumble for landscaping purposes, but restrictions may apply in a given area, so check online before your trip to ensure you are allowed to mine these rocks.

Specific Tools

Creek and river collecting don't usually require many tools. The most important items you'll need are the proper pieces of attire, especially good boots for walking along slick gravel bars and bedrock. A sturdy pair of rubber boots or even hip waders come in handy for crossing slow and shallow waterways. A backpack is also helpful when walking creeks and rivers. Buckets work but are a bit of a burden if you have to navigate around or through overgrown foliage or cross a tricky part of the waterway.

A geology pick and/or gad can come in handy to pry stubborn specimens that are lodged in between other rocks or stuck in hard, compacted sand. Sometimes you'll need to break open rocks to see what's inside or to release specimens stuck in matrix. A small crack hammer works well for opening rocks but also adds to the weight you'll be carrying around.

A sturdy gem scoop can really help when creek and river hunting. It saves your back the strain of constantly bending over and can allow you to reach specimens without getting wet. A gem scoop also helps you to balance on wobbly round rocks and when in the actual creek or river itself.

Best Times

The best time to hunt along rivers and creeks is when the water is low enough to start exposing gravel bars. These times can vary depending on how the waterway is fed and if anything is blocking its way downstream. Generally, warmer months and after snowmelt has slowed are the best times, but anytime the water gets low is a great opportunity. Many useful websites and apps track river levels, and keeping an eye on these can considerably improve your chances of finding gravel.

Rockhounding in Oceans and Lakes

The gravel in oceans and lakes quite often holds mineralogical riches. Almost everybody loves a day at the beach—especially when there is treasure! Beach agates in particular are highly sought after and have gotten many people hooked on rockhounding. Jasper, quartz, fossils, zeolites, beach glass, and glass floats are just some of the many other prizes a rockhound can find along the beaches of North America.

Using Maps

Using maps to plan a trip to a lakefront or oceanfront is pretty straightforward: Just use a printed or online map to pick a beach you are legally able to access. Of course, even if you are allowed to access a site physically, you may not be allowed to pick up and take rocks and minerals from the site. This is the case in some provincial, federal, and state parks. Call ahead to make sure it's okay to rockhound if you're not sure of the policies in a certain location.

Finding the Best Place to Stop

Finding the best place to stop is going to depend on what exactly you are looking for. Most rockhounds go to the beach to look for agates, jasper, and other hard minerals found in gravel deposits at these waterways. Lakefronts are particularly full of gravel deposits.

The ocean is a bit trickier when it comes to gravel deposits, as the changing tides and intense waves can move gravel unpredictably. A beach may be covered with gravel one day and completely sanded the next. Because less durable minerals or softer rocks and fossils are easily abraded and worn, they are rarely found on beaches; most are found in or near exposures of the formations in which they originated, where they are not subject to water abrasion and wear.

When you're planning a trip to the beach, pick an accessible spot you want to go to. Then note a few beaches nearby in case your first, second, or even third choices don't pan out. This is also a good plan of attack if you find a good spot that every other rockhound is drawn to as well. It can pay off to move a few miles down the shoreline.

What to Look For

If you're looking to find beach agates or other minerals hard enough to survive the unforgiving waves of the ocean, the very first thing you're going to need to find is gravel. Once you find gravel, there are a few different modes of attack. If the tide is out and there is gravel everywhere, take your time slowly walking along the beach, keeping your eyes focused immediately in front of you to find treasures.

If the tide is in the middle of going out or coming in and is still covering and tumbling the gravel—and you're a bit of a daredevil—you can do what is known as chasing waves. The idea is that as a wave recedes, you race to rockhound the freshly mixed gravel before the next wave comes back in and mixes the gravel up again.

CAUTION

Chasing waves can be a somewhat risky method of rockhounding on the beach and isn't recommended for everyone. Sneakier waves can roll in without warning and suddenly you may find yourself in the ocean. There can also be large debris in the water that can be dangerous.

If the tide is high, there won't be much of any rock or fossil to find, as most of it will be underwater. Always check tide charts before a trip to the ocean.

Deposits along lakes are easier to rockhound, as the water is usually more merciful. The Great Lakes are known for a wide variety of collectible minerals, including Lake Superior agate, Petoskey stone, fluorescent sodalite, crinoids, and chlorastrolite, as well as a rock known as puddingstone, which is a blend of quartzite and red jasper.

Gold hunters look for sand, specifically black sand that may have gold or other precious metals in it. Some people will collect sand in vials to search for microscopic garnets and other tiny crystals in it once at home.

Specific Tools

For general gravel rockhounding at a beach, you won't need much in the way of tools. Something to put what you find in is important, especially if you plan on bringing a lot of material home. A sturdy bag, backpack, or plastic bucket will serve well.

A geology pick and/or gad can come in handy to pry stubborn specimens that are lodged in between other rocks or stuck in hard, compacted sand. Sometimes you'll need to break open rocks to see what's inside or to release specimens stuck in matrix. A small crack hammer will be helpful for busting open rocks in these cases, but keep in mind that it also adds to the weight you'll be carrying around.

A gem scoop is very handy to scoop up rocks without straining your back. There's a lot of bending over when beach rockhounding, and a gem scoop is a lifesaver, especially on those days when there's gravel as far as the eye can see.

Best Times

For ocean rockhounding, the best time to look for gravel is in the winter months, especially after big storms that rip the sand off beaches and expose gravel. That's not to say that you can't find gravel in the summertime, but it is less likely. Be sure to check the tides, as they will also affect the day and time you plan to visit the beach. The best time to rockhound an ocean beach is when the tide is going out, especially if it's a negative low tide (the lowest tide of the day).

When lake rockhounding, start checking sites in early spring, when ice, snowmelt, and the water recede, exposing gravel. As waters continue to go down, keep checking sites. Gravel will continue to be uncovered by the water.

Rockhounding in Mountains and Forests

Have you ever heard the phrase "There's gold in them thar hills"? While this is true—there *is* gold in "them thar" hills—the mountains and forests of North America are full of many other collectible rocks and minerals as well. Quartz crystals, smoky quartz, tourmaline, and many other crystal specimens are often found in mountainous regions. Emeralds were discovered in Canada's Yukon Territory in 1998. Columnar basalt used in landscaping is also found in mountainous regions, but make sure you check collecting regulations in the area you are planning to mine for this rock, as restrictions may apply. Fossils can also be found in mountains and forests, always near exposures of such sedimentary rocks as sandstone, limestone, and shale. Fossils of the pelecypod *Inoceramus* are abundant in the forests and plains of the central United States and south-central Canada.

Using Maps

Physical maps are essential when rockhounding in mountains and forests. It's easy to get turned around while traveling a maze of long and winding trails or wandering through the trees, and cell phones often lose reception. A compass is helpful as well.

Finding the Best Place to Stop

Mountain creeks and rivers are excellent places to find material that has been weathered out of the mountainsides and deposited in gravel bars downstream. Waterways are one great way to find the source of the material. Follow the creek or river upstream until you stop finding the material you're looking for. Then start looking more closely around the area where you last saw the material. The source is likely here.

ROCKHOUND IN THE KNOW

Following creeks and rivers to their source is how many gold miners have located gold deposits.

When exploring mountains and forests, look for exposed rock outcrops or areas of erosion. In the mountains, just above the tree line is the easiest area to find exposed rock. Rockhounding below the tree line is a bit more difficult, because much of the local rock is covered by thick layers of dirt and plants.

Keep an eye out for rocks in the soil. If you start finding material, try to figure out if you're standing on the source or if it was moved there long ago. Rocks moved by rivers or glaciers will be well-rounded, whereas material closer to the source will be much rougher. Glaciers are responsible for depositing many minerals, such as sought-after agates, throughout North America. Noting where ancient rivers or glaciers once were can help you find potential deposits of specimens.

What to Look For

While a bit hidden, many treasures await the rockhound high in the mountains and deep in the forests. With volcanic activity comes many minerals. Many famous quartz crystal deposits in North America are found high up on mountaintops. The mountain ranges of western North America are known for their fine quartz crystals, agates, geodes, jades, garnets, zeolites, and sapphires. The mountains in Colorado are known for their fine specimens of aquamarine, fluorite, smoky quartz, and topaz. The Great Smoky Mountains in western North Carolina are home to many ruby deposits.

Forests are prime sources of beautiful specimens. Amazonite is found in the forests of Colorado, copper nuggets in the forests of Wisconsin and Michigan, opal in the forests of Louisiana, and fluorite in the forests of Kentucky.

Look for unusual colors, shapes, textures, and landforms. Certain minerals, such as blue azurite and red cinnabar, stand out from the drab colors of common rocks and gravel. And unlike the rounded and ill-defined shapes of common rocks, crystals have straight edges, symmetrical terminations, and smooth, lustrous faces. Fossils, whether those of leaves or nautiloids, always have distinctive and recognizable shapes and patterns.

Unusual landforms can also indicate the presence of collectible minerals and fossils. Rock outcrops, places where subterranean rock formations are exposed on the surface, often contain minerals and fossils that cannot be found elsewhere.

Specific Tools

The tools used for mountain and forest rockhounding depend on what you're looking for and the site. For some locations, it's just a matter of getting to the area and picking material up off the ground—or maybe doing a bit of light digging and screening. In other areas, you may need hard-rock mining equipment to free specimens from their matrix. Do your research on the material you plan on rockhounding, and bring the appropriate tools for the job.

Backpacks tend to be better for carrying specimens in mountains and forests, rather than a 5-gallon bucket. It's much easier to walk in steep or overgrown areas with your hands free.

Best Times

Mountain rockhounding is usually only done after the snow melts. In some places that never happens. Check on snow levels before traveling to any high-altitude sites, as the snow can completely cover some rock-hounding areas even in summer.

Many heavily forested areas aren't much fun to collect in until the rainy season is over. (For most people, mud and cold rain are not comfortable to navigate.)

The best time to collect in both mountains and forests is late spring through early autumn. Avoid overexposure to the sun when at high-altitude sites with little shade.

Rockhounding in Deserts

The deserts of North America hold mineralogical treasures in their beautiful and foreboding landscape. Many coveted finds have inspired people to leave their city lives behind to dig and mine for gems, gold, and fossils in some of the most inhospitable terrain on the continent. Many minerals such as sunstone, malachite, azurite, and variscite are known for being found in these arid climates, though the sky is the limit when it comes to desert minerals. Many fossils, especially petrified wood, can also be found throughout North American deserts. Rocks such as sandstone and shale are also easily found here.

Using Maps

Desert rockhounding will take you deep into the middle of nowhere and down some of the quietest yet most beautiful highways and dirt roads you've ever seen. It's easy to get lost. Bring a printed map and compass, and have a dependable app set up on your phone. Mark the areas you want to visit, but be open to finding new sites, as both printed and online maps don't always show how tricky roads can be to navigate, especially those in the desert.

Finding the Best Place to Stop

In the desert, look for places of erosion, which will have more exposed rocks. Look at the landscape and think like water: If you were a body of water on the landscape in front of you, where would you flow? Dry creek beds are a good start, but also think of where the rare rainfall is going to flow down hillsides. Follow these potential water paths, keeping a close eye out for any possible treasures. Once you find good float material, you can either work on finding the source of said material and dig or continue exploring the desert more broadly for treasure.

What to Look For

In the northwestern United States, rockhounds comb the high deserts for many colors and variations of agate, jasper, petrified wood, and obsidian. Oregon is home to Oregon sunstone, a variety of copper-included labradorite feldspar that is only found in Oregon. The southwestern US deserts are known for their copper-bearing gemstones, including turquoise and variscite. The northwest corner of Nevada is home to some of the most beautiful opal in the world. World-class tourmaline crystals come from San Diego County in California. While there are claims on many of these minerals, there is a lot of room to roam in the desert, and the adventurous rockhound is sure to find some treasure. There are also many claimed operations that offer fee-dig services, giving everyone the chance to find beautiful rare minerals.

Specific Tools

There are two modes of attack when desert rockhounding: Wander around and pick up float material lying on the ground, or do a lot of digging. If you plan on wandering, bring a good backpack or bucket to carry your treasures in. A geology pick, Estwing Paleo Pick, or gad and a small collapsible shovel are convenient for removing what many rockhounds refer to as icebergs from the desert soil.

Digging always involves a shovel, of course, but having a pick, screens, an array of hammers, and some chisels is a good idea as well. Some minerals such as agates and Oregon sunstone may just need to be dug up and screened out of the dirt. Others may require quite a bit more elbow grease to remove from the host rock. Many agate, geode, and thunder egg deposits have to be freed from hard host minerals such as basalt and rhyolite. Research what you plan on collecting and bring the recommended tools for the job.

ROCKHOUND IN THE KNOW

What is an "iceberg" in the rockhounding world? Often only the tip of a mineral can be seen poking out of the desert floor, while a huge chunk lies below the surface—just as icebergs appear in the water. Many of the best specimens in my own collection were found in these desert "icebergs."

Best Times

The desert is usually a harsh environment at any time of year. During the summer, high temperatures make it impossible simply to wander around, let alone dig a hole or remove minerals from hard host rock. The nights, even in the summer, can be the opposite, with temperatures plummeting to near freezing. The high desert even gets snow in the winter. Plan your desert trips for the autumn to avoid blistering heat. If possible, keep your rockhounding to the daytime to avoid cold nighttime temperatures.

Packing and Transporting Your Treasure

It is important to plan ahead how to pack and transport your material from the field to your vehicle to your home in the safest manner possible.

You didn't spend all that time rockhounding just to break everything, right? Luckily, most minerals the average rockhound collects can simply be put into a bucket or bag and will not suffer much—if any—damage in transport. Agate, jasper, and petrified wood are all examples of minerals that can survive a bucket.

For soft and fragile minerals—generally, crystal clusters of some type—you'll need to wrap them in something. Paper towels, newspaper, Bubble Wrap, tissue paper, and toilet paper are a few go-to wrappings rockhounds commonly use. Some people save egg cartons to store wrapped delicate specimens.

Do your research on your finds to ensure they get home safe and sound—and ready for the next step: preparing for display, jewelry, and more! In the next chapter, you'll learn how exactly to clean and polish and/or cut the treasures you've brought back from your adventures to show them off to the world (or just friends and family).

Chapter 4

PREPARING AND SHOWING OFF YOUR FINDS

Now you have some rocks and minerals (or maybe a few bucketfuls!). What's next? Half of the fun of being a rockhound is what you do with the treasures you find out in the field. For some, simply putting a found rock in a garden or displaying a fossil on a bookshelf is all they need to enjoy their bounty. For others, cutting and/or polishing their own minerals to wear as a piece of jewelry is the ultimate satisfaction. In the following chapter, you'll learn how to prep your own finds for whatever you choose. From cleaning and displaying as found to polishing, cutting, and more, this chapter will tell you how to share your treasures with the world—or simply enjoy them in the comfort of your own home.

How to Clean Your Finds

It's no surprise that most of the minerals you find will be dirty, from a light to heavy covering in soil or clay to iron stains. You'll start simply by cleaning your haul.

Most of the rocks and minerals you bring home can be scrubbed with warm water; a mild soap, such as hand soap or dishwashing soap; and some sort of brush. Old toothbrushes work great as rock brushes. For stubborn pockets of clay, scrub the specimen with soap and water, let it dry out completely, then repeat scrubbing and drying as necessary.

ROCKHOUND IN THE KNOW

A pump spray bottle can help clean off stubborn clay. For large specimens, a pressure washer can power off the grime. Start soft and work your way up to a more powerful tool, especially with material that is new to you. You don't want to start with a pressure washer and destroy a beautiful specimen.

For stubborn staining and mineral crust such as calcium deposits, soak the mineral in an acidic solution. The most commonly used acids for cleaning minerals are muriatic acid, oxalic acid, and vinegar. Some people use ketchup to clean iron stains off rocks, but it can be a pricey way to clean minerals.

When removing stains, first determine just how bad the staining is, and then decide how strong a solution you'll need. Deeper stains will need a more acidic solution. Soak rocks in a bucket with acid and water, and place it outside the home and out of reach of children and pets. If you're using vinegar, a large Mason jar with a lid could be used inside the home. How long the soaking takes depends on how stained the rocks are, as well the material of the stain, how strong your acid solution is, and the weather. Acids work better when it they're warm. Using a

Crock-Pot outdoors that is designated for cleaning minerals can massively speed up the cleaning process.

How to Prepare Your Finds

Once cleaned, some minerals will need to be prepped a bit to show them best for display. Prepping may consist of cutting the matrix or mineral itself on the bottom so it can stand alone. You may need to repair a broken crystal from a cluster or as a specimen itself. Many quartz and other crystal clusters have broken crystals that are glued back onto the cluster to create a more aesthetically pleasing specimen.

Many fossils will need to be prepped with a machine such as an air scribe to expose the fossil from the host rock it's trapped in. Fossils will often break when prepping, and gluing pieces back together is common practice.

Some mineral matrices and crystals require stabilization with various glues and vacuum treatments. One popular method is to lightly coat specimens with a water-clear epoxy. If left unstabilized, these minerals may deteriorate over time, usually due to oxidization.

ROCKHOUND IN THE KNOW

Much of the turquoise collected is not stable enough when untreated to be used in jewelry. The turquoise is stabilized to strengthen it so it can be cut and set in necklaces, rings, and more.

Many agates such as Lake Superior agates and Fairburn agates are soaked and heated with mineral oil to bring out the natural colors and patterns found in the stone. It's a way of polishing an agate without tumbling it. Collectors of Superiors and Fairburns do not often cut or machine-polish their specimens but prefer to retain the shape in which they were found. This method can also be used with other cryptocrystalline silicates such as jasper, flint, and chert.

The Lapidary Process

Lapidary is the term that refers to both the craft of cutting and polishing rocks and minerals and to the person that does the work. The lapidary process may follow cleaning and prepping, unless the rockhound has decided to simply display their specimen.

There are many forms of lapidary work, and the following sections cover the most popular methods of the craft. As you learn about these methods, keep in mind that there are many practices that need to be learned to form a solid base to the craft, but nothing is set in stone. The sky is the limit when it comes to cutting and polishing your rocks and minerals. Practice, practice, practice, and find out what works for you.

Polishing

There are as many techniques as there are different machines to polish and shape rocks and minerals. They usually consist of some sort of silicon carbide, steel, or diamond-infused wheel; sandpaper; or flat lapidary disks. You'll start with a coarse abrasive, usually made from silicon carbide, called grit to shape the stone, and then progressively work your way through finer and finer grits. Finally, you'll move to polishing on a soft buffing wheel coated with a polishing compound.

Tumbling

Tumble-polishing is one of the first techniques most beginner rockhounds learn for polishing rocks and minerals. The machines are relatively inexpensive, and the labor involved is low.

The best thing you will learn from tumbling is patience. Most material you will tumble will take a month minimum from beginning to end.

Tumblers come in two varieties: rotary and vibratory. Rotary tumblers consist of a barrel that is "charged" with rocks, water, abrasive grit, and a polish in the final stage. The barrel is sealed and placed on a motorized machine that turns the barrel and tumbles the material inside. Vibratory tumblers vibrate material rather than rotate it. They also use various stages of grit and polish, but don't change the shape or round out hard material nearly as much as rotary machines. They are often used for softer minerals for this reason. You can check the contents of a vibratory tumbler much easier than a rotary tumbler.

There are four to five stages involved in tumble-polishing rocks and minerals. The first three stages consist of using a series of abrasives consisting of a coarse grit, a medium grit, and a fine grit. The coarse grit is used to shape the stones. If the material you put in the barrel is round and shaped (e.g., beach agates), then you won't have to do much shaping. Some people skip this stage when tumbling rocks. The next two grits start to polish out the scratches left after the first stage. After the abrasives, you move into polishing. Powders such as tin and cerium oxides are added to the machine to polish the stones during this stage. After the polishing stage, some people do what is called a burnishing stage, usually using borax or other powdered detergents. This stage can remove residue that may have built up during the polishing stage. It can also add more luster and shine to your stones for a glassy finish.

Cutting

To cut most material, you will need a diamond-blade saw. The two types of saws found in the shops of most lapidaries are the slab saw and the trim saw. Both saws are made with a diamond blade for precise cutting.

ROCKHOUND IN THE KNOW

A diamond blade will not cut the skin. Think of it as more like a spinning nail file than an actual blade.

The typical slab saw is an enclosed machine that consists of the diamond blade, a clamp, some sort of cooling liquid, and an automatic feed drive. It is used to cut slabs of rocks, but it is also used for cutting faces into stones that will eventually be polished for a display piece. Slab saws are also used to cut stones into a round geometrical shape that will then be used in a sphere-making machine. Slab saw blades are generally 10"–36" long and vary in thickness. The stone being cut should usually be no more than one-third the diameter of the blade.

Trim saws are similar to slab saws, except the unit is open and there is usually no clamp. Trim saw blades are generally 4"–8" in diameter and come in a variety of thicknesses. These saws are mostly used to cut preforms (roughly shaped cabochons) from slabs that will be used to make cabochons. They can also be used to cut smaller stones, such as agate nodules and thunder eggs, in half. They can also be used to trim, shape, and prep stones in matrix.

Cabochons

A cabochon is a cut and polished gemstone with a domed top and usually a flat bottom. They are cut into a wide variety of shapes—most commonly ovals, rectangles, and circles, but almost any shape is possible. Rock shows are great places to see a wide range of shapes made by local lapidaries.

To make a cabochon, use a slab saw to cut a slab of the material you want to make it out of. Once you have a slab, pick a shape and which part of the slab you want to cut it from. A properly placed shape can make a cabochon a real work of art. After you have picked and marked the part of the stone you are cutting, cut a rough shape around it to make a preform. Then use a series of wheels with grit, called a cab arbor, to shape and polish the stone.

Spheres

Turning rocks and minerals into spheres is a popular lapidary craft. Sphere-making machines and slab saws are required to make the spheres. The lapidary must first make twenty-two cuts in slab with a slab saw to produce an octahedron. The octahedron then goes into the sphere maker. The sphere maker applies water and abrasive grit to rotating or gyrating cups that are pressed against the already rounded specimen.

Faceting

Faceting is a process of grinding and polishing multiple planes on a gemstone. Faceted gemstones are most often used in jewelry but are also of interest to mineral collectors. Traditionally, hard, transparent stones with a high refractive index are faceted, as they sparkle the most and are tough enough to be worn as jewelry. Soft, translucent, or opaque stones are faceted for collectors, but not used in jewelry.

Storage, Display, and Artisanal Uses

Once your rocks and minerals have been cleaned, prepped, and polished and/or cut if desired, there are endless ways to store, display, and use them. The sky is the limit; find what storage methods, display methods, or artisanal uses work for you.

Storing Specimens

Storage of a collection depends on what you're collecting. Soft and fragile minerals will need special attention to keep from getting damaged. Harder material can be kept in buckets in the backyard and never really get damaged.

Plastic 5-gallon buckets are the quintessential apparatus when it comes to storing large amounts of harder specimens. Milk crates and wooden boxes are also popular storage units. If you plan on storing specimens outside uncovered in 5-gallon buckets, it's wise to drill a few holes in the bottom to allow rainwater to drain out. Algae and the buildup of fallen leaves and other organic matter cause a real mess in a bucket.

Displaying Specimens

Fancy lighted cases with acrylic stands and labels, repurposed vintage drawers, a simple bookshelf—there are endless options for how to display your rocks and minerals. The most important aspect of displaying your collection is simple: a shelf or somewhere to place your specimen. Then take into consideration how your specimen sits. Can it sit on its own or might it need a stand?

There are many options of acrylic, metal, and wood stands made specifically for rocks, gems, minerals, and fossils. You can find them at your local rock shop or online. Think outside of the box and find what works for you.

FROM ONE ROCKHOUND TO ANOTHER

I once saw a mineral sphere display with every orb set on galvanized piping and couplings. It was an absolutely beautiful display of different materials, colors, and color patterns. It goes to show that a little creative ingenuity at your local hardware store can make for a professional-looking display.

Labeling is important for any collection, especially when you start accumulating a lot of rocks or minerals. It's all too easy to forget exactly where something came from or even what it is. You can print labels with the name of the mineral, the chemical composition, where it was found, and any other information for each specimen on display; keep a list in a journal; take photographs and write notes on the back; or even start an *Instagram* page to keep track of what you have found and where.

Using Specimens in Jewelry, Arts, and Crafts

Beyond putting a specimen on a shelf or turning it into a piece of jewelry, there are many options to turn your treasures into something that expresses your creativity. The following are great ideas for crafty things you can do with your rocks and minerals:

- **Mosaics:** With a frame, backboard, and glue, tiny specimens of minerals and gemstones can be arranged into patterns by color, shape, and luster, and fixed into attractive mosaics for wall hangings.
- **Paperweights:** Glue a custom-cut piece of felt onto the flat surface of medium-sized specimens of rocks and minerals to transform them into attractive paperweights.
- **Bookends:** Glue a custom-cut piece of felt onto the flat base of large, cabinet-sized specimens to make unusual and functional bookends.
- **Creative Mountings:** Many unusually shaped objects of wood, plastic, or metal can make striking mounts for showing off mineral specimens. For example, large crystals can be especially eye-catching when mounted in twisted pieces of driftwood. Spice racks can serve as shadowboxes for displaying multiple specimens.

- **Stone Carvings:** Many softer rocks and minerals, such as soapstone, marble, certain forms of gypsum, calcite, and limestone make easily workable and attractive mediums for stone carving.
- **Paintings:** Smoothed and rounded river and beach rocks make an unusual "canvas" for painting.
- **Backlit Displays:** Place small lights behind transparent or translucent crystals to make unusually dramatic displays.
- **Rocks in Jars:** Fill transparent glass jars, vases, and bottles with rounded river or beach stones that have been tumbled or painted; use your imagination to create color patterns and designs.

There are myriad ways to enjoy and share your rocks, gems and minerals, and fossils—use your imagination!

Where to Start

Now that you have discovered more about rockhounding, explored the best tools for the job, learned where to look in whatever setting you choose to rockhound, and uncovered the different ways to clean, prep, polish, and display your finds, it's time to get out in the field. Of course, it can feel overwhelming when you first get started: so many rocks, gems, and more to find, and so many places to find them! But don't fear; Part 2 will help you kick-start your adventures. In the pages that follow, you will find 150 profiles for popular rocks, gems, minerals/mineraloids, and fossils, along with all the details you need to find and collect them. Flip through to the profiles that stand out to you the most, and plan your rockhounding trips around those rocks and minerals. You can also use this section to identify any unexpected specimens in the field, so be sure to bring it along on your quest.

PART 2

Rocks, Gems, Minerals, Mineraloids, and Fossils of North America

Y ou've explored the history of rockhounding and why it's such a popular hobby, learned more about the tools and equipment found in every good rockhound's toolbox, and uncovered the where and how of locating, collecting, and processing rocks, gems, minerals/mineraloids, and fossils. So it's time to get into the real nitty-gritty details—the specific treasures you can start hunting for today.

This part provides complete profiles for 150 popular rocks, gems, minerals/mineraloids, and fossils hidden across North America. You'll find information about how to identify each specimen (including a color photograph of what it looks like in the field), what regions and geological settings of Canada and the United States it is located in, and what purposes it is commonly collected for. Each profile also offers helpful tips for successfully collecting and processing your prizes like a pro. And for even more information on rockhounding, from additional guides to popular rock shops and shows, be sure to check out the resources section in the Appendix at the back of this book.

Defining Headers and Terms

Before you dive into the specific rock, gem, mineral, mineraloid, and fossil profiles that follow, you'll want to take a closer look at the common identification terms you will discover here. Each rockhounding treasure has certain classifications, from the major and minor minerals of a rock to the hardness rating and crystal habit of a gem or mineral to the fracture of a mineraloid. The following sections cover what each of these terms means, so you can make the most of the profiles. (For a full list of rockhounding terms and definitions, see the Glossary at the back of this book.)

Rocks

- **Rock Type:** Refers to the three basic types of rock: igneous (formed when hot material cools and solidifies), metamorphic (formed when a rock changes its original form due to immense heat or pressure), and sedimentary (formed from pieces of other rock or organic material).
- **Structure:** Refers to such visible features as bedding (layering), flow banding, the presence of vesicles (hollow voids), and uniformity or lack of uniformity of crystalline mineral components.
- **Major Minerals:** Also known as essential minerals, these are the major mineral components of a rock that are used to classify the rock.
- **Minor Minerals:** Also known as accessory minerals, these are present in lesser amounts and are not used to classify the rock.
- **Fossils:** Any remains, impressions, or traces of past life-forms that are preserved in a rock environment.
- **Color(s):** The color (or combination of colors) exhibited by a rock.

- **Texture:** The surface appearance of a rock as determined by grain size; may be glassy (without visible grains), fine-grained, medium-grained, or coarse-grained.
- **Description:** A brief description and history of the rock.
- **North American Locations:** Geological settings and regions in Canada and/or the United States where you can find the rock.
- **Collecting Tips:** Information on how best to go about finding the rock when out in the field, including what to look for and where the rock is most commonly found in the previously detailed locations.
- **Processing Tips:** Information on how to clean and display the rock, as well as any popular lapidary processes and uses.

Gems and Minerals

- **Category:** The category of a mineral is determined by its chemical composition as defined in the Dana System of Mineralogy.
- **Crystal System:** A seven-category classification of gems/minerals based on their crystal formation. Each crystal system is defined by its number of crystal axes, the lengths of these axes, and the angles at which these axes intersect.
- **Crystal Habit:** The external shape of a gem/mineral crystal or group of crystals.
- **Cleavage:** The tendency of a mineral to break along one or more regular, smooth surfaces that are related to crystal structure.
- **Fracture:** The appearance of the surface of a freshly broken mineral; may be uneven, conchoidal, hackly, earthy, etc.
- **Mohs Scale:** A 1–10 scale of gem/mineral hardness, 1 being the softest (talc) and 10 being the hardest (diamond).
- **Color(s):** The color(s) or combination of colors exhibited by a gem/mineral.

- **Luster:** A reference to the shine of the gem/mineral.
- **Streak:** The color of the gem/mineral when in powder form. A "streak test" is accomplished by rubbing the gem/mineral against an unpolished piece of porcelain.
- **Transparency:** A reference to the amount of light that can pass through the gem/mineral.
- **Description:** A brief description, history, and name origin of the gem/mineral.
- **North American Locations:** Geological settings and regions in Canada and/or the United States where you can find the gem/mineral. The listed states and provinces indicate only where better specimens have been collected. If a state or province is not listed, it does not necessarily mean that a mineral cannot be found there.
- **Collecting Tips:** Information on how to best go about finding the gem/mineral when out in the field, including what to look for and where it is most commonly found within the previously detailed locations.
- **Processing Tips:** Information on how to clean and display the gem/mineral, as well as any popular lapidary processes and uses.

Mineraloids

- **Category:** The category of a mineraloid is determined by its amorphous structure and indefinite chemical composition.
- **Crystal Habit:** The external shape of a mineraloid (amorphous or massive).
- **Fracture:** The appearance of the surface of a freshly broken mineraloid; may be uneven, conchoidal, hackly, earthy, etc.
- **Mohs Scale:** A 1–10 scale of mineraloid hardness, 1 being the softest (talc) and 10 being the hardest (diamond).

- **Color(s):** The color(s) or combination of colors exhibited by a mineraloid.
- **Luster:** A reference to the shine of the mineraloid.
- **Streak:** The color of the mineraloid when in powder form. A "streak test" is accomplished by rubbing the gem/mineral against an unpolished piece of porcelain.
- **Transparency:** A reference to the amount of light that can pass through the gem/mineral.
- **Description:** A brief description, history, and name origin of the mineraloid.
- **North American Locations:** Geological settings and regions in Canada and/or the United States where you can find the mineraloid. The listed states and provinces indicate only where better specimens have been collected. If a state or province is not listed, it does not necessarily mean that a mineraloid cannot be found there.
- **Collecting Tips:** Information on how to best go about finding the mineraloid when out in the field, including what to look for and where it is most commonly found within the previously detailed locations.
- **Processing Tips:** Information on how to clean and display the mineraloid, as well as any popular lapidary processes and uses.

Fossils

- **Category:** The category of a fossil is determined by its chemical composition and the life-form remains that they preserve.
- **Mohs Scale:** A 1–10 scale of hardness, 1 being the softest (talc) and 10 being the hardest (diamond).
- **Color(s):** The color(s) or combination of colors exhibited by a fossil.

- **Luster:** A reference to the shine of the fossil.
- **Transparency:** A reference to the amount of light that can pass through the fossil.
- **Description:** A brief description and history of the fossil.
- **North American Locations:** Geological settings and regions in Canada and/or the United States where you can find the fossil.
- **Collecting Tips:** Information on how to best go about finding the fossil when out in the field, including what to look for and where the fossil is most commonly found within the previously detailed locations.
- **Processing Tips:** Information on how to clean and display the fossil, as well as any popular lapidary processes and uses.

AEGIRINE

Category: Inosilicate
Crystal System: Monoclinic
Crystal Habit: Prismatic
Cleavage: Good in one direction
Fracture: Uneven
Mohs Scale: 6

Colors: Dark green, reddish brown, black
Luster: Vitreous to slightly resinous
Streak: Yellow gray
Transparency: Translucent, opaque

Description: Aegirine is a member of the clinopyroxene group of inosilicate minerals. It derives its name from the mythological Norse god of the sea, Ægir. Aegirine is also known as acmite.

North American Locations: Canada: Mountains in Ontario and Quebec. United States: Mountains in Arizona, California, Colorado, Massachusetts, and North Carolina.

Collecting Tips: Aegirine is commonly found in mafic igneous rocks like syenites and syenitic pegmatites. It can also be found in metamorphosed iron-rich sediments, schist, and fluid-enhanced metamorphic rocks. Aegirine is easily identifiable by its dark green monoclinic, prismatic crystals.

Processing Tips: Aegirine is collected mainly as a crystal specimen. Dish soap and water can be used to remove dirt. The crystal clusters can be difficult to clean and prep if highly stained; do some research online for the best methods to clean stubborn specimens. Solid specimens can be cut as cabochons or faceted.

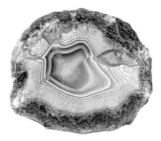

AGATE

Category: Tectosilicate
Crystal System: Hexagonal
Crystal Habit: Cryptocrystalline, massive
Cleavage: None
Fracture: Conchoidal

Mohs Scale: 6.5–7
Colors: Brown, white, red, gray, pink, black, yellow
Luster: Vitreous to waxy
Streak: White
Transparency: Translucent, opaque

Description: Agate is a translucent variety of microcrystalline quartz. It got its name from Theophrastus, a Greek philosopher, who named it after the Achates River (now called Dirillo), where it has been collected for over three thousand years.

North American Locations: Along ocean and lake shorelines, and in river gravels, plains, mountains, and deserts across North America.

Collecting Tips: Agates can be commonly found as an alluvial deposit and in the gravel of shorelines, riverbanks, and creek banks. Agate can also be found in cavities of igneous rocks. The sun helps agates glow; rain helps agates shine as though polished, especially in areas with lots of gravel.

Processing Tips: Agates are a popular tumbler material. Specimens with banding are often cut to expose the banding on a polished, flat face. Agate is also often cut into slabs, which can be polished and displayed or used for cabochons.

| Mineral | **ALBITE** |

Category: Tectosilicate
Crystal System: Triclinic
Crystal Habit: Tabular
Cleavage: Good in two directions
Fracture: Uneven
Mohs Scale: 6–6.5

Colors: White, colorless, yellow, pink, green
Luster: Vitreous to pearly
Streak: White
Transparency: Translucent, transparent

Description: Albite is the end member of the plagioclase-feldspar series. It derives its name from the Latin *albus*, meaning "white." It is widely used in glass and ceramics as both binding materials and glazes.

North American Locations: All geological settings across North America.

Collecting Tips: Albite occurs in pegmatites, felsic and igneous rocks, and low-grade metamorphic rocks. It can also form through chemical deposition in sedimentary environments. Albite crystals are easily identified by their white color and crystal habit. They are commonly found alongside biotite, hornblende, muscovite, orthoclase, and quartz.

Processing Tips: Most albite specimens are simply cleaned, possibly trimmed, and put on display. Stained specimens can be cleaned by soaking in oxalic acid. Solid crystals can be cut into cabochons. Transparent specimens can be faceted.

AMAZONITE

Category: Tectosilicate
Crystal System: Triclinic
Crystal Habit: Prismatic
Cleavage: Good in two directions
Fracture: Uneven
Mohs Scale: 6–6.5

Colors: Green, bluish green
Luster: Vitreous
Streak: White
Transparency: Translucent, opaque

Description: Amazonite is the blue-green variety of microcline feldspar. It got its name from the Amazon River, although it has not been found anywhere near this river. It has been collected for over two thousand years in Egypt and what was once Mesopotamia.

North American Locations: Canada: No known deposits. United States: Mountains in Colorado and forests of Virginia. The world's best specimens come from Colorado.

Collecting Tips: Amazonite is easily recognized by its distinctive blue-green color. Look for pegmatites in granitic rocks. Care must be taken not to damage crystals when extracting, although broken crystals and crystal clusters can be glued back together.

Processing Tips: Fine amazonite crystals are cleaned, sometimes trimmed, and displayed. Use sodium dithionite to clean stained specimens; stubborn staining can also be removed with oxalic or hydrochloric acid. Cleaning tips for specimens with multiple minerals can be found online. Amazonite can be tumbled and cut into cabochons for jewelry.

Mineraloid

AMBER

Category: Organic gemstone (hydrocarbon)
Crystal Habit: Amorphous
Fracture: Conchoidal
Mohs Scale: 2–2.5

Colors: Yellow, orange, red, brown, blue
Luster: Resinous
Streak: White
Transparency: Translucent, transparent

Description: Amber is a type of fossilized resin that predominately came from extinct coniferous trees. It is commonly found with or near lignite coal, which is also fossilized tree and plant matter. The Greeks called amber *electron*, which is where the modern word *electricity* comes from. Amber can produce an electrical charge when rubbed.

North American Locations: Canada: Beach and river gravels, and along riverbanks in Alberta and Manitoba. United States: Beach and river gravels, and along riverbanks in Arkansas, New Jersey, New Mexico, and Utah.

Collecting Tips: Amber is easily recognized by its bright color and plastic-like appearance. It is light enough to float in heavily salted water.

Processing Tips: Amber is very soft and can be shaped by hand with sandpaper. An old-school method of polishing amber is using a gritty toothpaste and a lint-free cloth. Amber is popular in jewelry making.

AMETHYST

Category: Silicate
Crystal System: Hexagonal
Crystal Habit: Pyramidal, prismatic
Cleavage: None
Fracture: Conchoidal

Mohs Scale: 7
Colors: Violet, dark purple
Luster: Vitreous, glassy
Streak: White
Transparency: Transparent, translucent

Description: Amethyst is the purple variety of crystalline quartz that obtains its hue from iron. Its name comes from the Greek *amethystos*, meaning "not drunk." The ancient Greeks believed that cups carved from amethyst would protect against the intoxicating effects of wine. Amethyst is the birthstone for February.

North American Locations: Mountains and deserts across North America. Notable collecting sites are located in Arizona, California, Colorado, New Hampshire, North Carolina, and South Dakota, and in British Columbia, Canada.

Collecting Tips: Amethyst crystals are most commonly found in vugs or empty pockets of igneous rock. Hard-rock mining tools are required to expose these pockets and remove the crystals. Amethyst can also be found in geodes, thunder eggs, and agate nodules and veins.

Processing Tips: Fine amethyst crystals and clusters are cleaned, sometimes trimmed, and displayed. Crystals can be cleaned with soap, water, and a stiff plastic brush. Any iron staining can be removed with oxalic or muriatic acid. Amethyst can also be cut en cabochon, faceted, carved, and tumbled.

AMMOLITE

Category: Fossil

Mohs Scale: 3.5–4

Colors: Violet, blue, green, yellow, red; all colors are iridescent

Luster: Vitreous, waxy

Transparency: Translucent to opaque

Description: Ammolite, a trade name, is a rare gem variety of Cretaceous ammonite fossil. The fossils are made mostly of aragonite but can contain various amounts of calcite, silica, pyrite, and other minerals. Colors are iridescent and range from green, blue, indigo, and violet to red, yellow, and orange.

North American Locations: Canada: Rivers and plains in Alberta. United States: No gem-quality deposits.

Collecting Tips: The ammolite-mining areas in Canada are in ancient river gravels and along the exposed banks of rivers. All known sites are on privately owned or leased tribal land and not open to unauthorized collecting. Rockhounds can sometimes gain access to mines through gem and mineral shows and clubs. Ammolite is easily recognized by its intense play of iridescent colors.

Processing Tips: Ammolite is soft, and care must be taken when polishing specimens. Finished material is often sealed with a polyurethane or epoxy to protect it from scratches and drying out.

AMMONOID

Category: Fossil
Mohs Scale: 3–4
Colors: Gray, tan, brown, brownish black

Luster: Earthy, dull
Transparency: Opaque

Description: Ammonoids are an extinct group of cephalopods that are now only found as fossil evidence of ancient seas. Ammonoids became extinct at the end of the Cretaceous period when the dinosaurs also disappeared.

North American Locations: Along riverbanks and lake shorelines, and in mountains, plains, and deserts across much of North America.

Collecting Tips: These fossils are most easily recognized by their spiral-shaped shells divided into chambers. Care should be taken when extracting fossil specimens as they can be damaged or broken easily. Broken specimens can be glued back together. Packing material such as newspaper or Bubble Wrap is helpful for getting specimens home undamaged.

Processing Tips: Many fossils of ammonites can be displayed as they were found in nature. Specialty tools can be used to help expose fossils from the host rock for a better display piece. Care must be taken when prepping fossils as they break easily.

ANALCIME

Category: Zeolite
Crystal System: Triclinic
Crystal Habit: Trapezohedrons, massive, granular
Cleavage: None
Fracture: Uneven

Mohs Scale: 5–5.5
Colors: White, colorless, gray, pink, greenish, yellowish
Luster: Vitreous
Streak: White
Transparency: Transparent, translucent

Description: Analcime was once considered a feldspar mineral but is now grouped with zeolite minerals. Its name is derived from the Greek *analkis*, meaning "weak," as it emits a weak electrical charge when heated or rubbed. Analcime is mined commercially to make silica gel.

North American Locations: Mountains across North America. Fine specimens have come from California, New Jersey, Oregon, and Washington, and from British Columbia and Quebec, Canada.

Collecting Tips: Analcime occurs primarily in vugs in basalt and mafic igneous rocks. It is recognized in the field by its white color and trapezohedral crystal habit. Hammers will be needed to break out or trim down specimens. Take care when extracting crystals from the host rock as they can be damaged easily.

Processing Tips: Analcime is collected mostly as a crystal specimen for display. Clusters of crystals are often left in the host rock and trimmed. Rare large, colorless crystals can be faceted or turned into cabochons.

ANDALUSITE

Category: Nesosilicate
Crystal System: Orthorhombic
Crystal Habit: Prismatic, acicular
Cleavage: Good in two directions
Fracture: Uneven
Mohs Scale: 7–7.5

Colors: Pink, reddish brown, yellow, green, white, gray, violet
Luster: Vitreous, subvitreous, greasy
Streak: White
Transparency: Transparent, translucent, opaque

Description: Andalusite was named for its presumed place of discovery: Andalusia, Spain. It turned out, however, that it was actually from the province of Guadalajara. Industrially, andalusite is used as a refractory in furnaces and kilns.

North American Locations: Deserts, mountains, and river gravels across North America. Fine specimens come from California and Massachusetts, and from Nova Scotia and Ontario, Canada.

Collecting Tips: Andalusite is found mostly in low-grade and regional metamorphic rocks but can sometimes be found in granitic pegmatites. The crystals are identified by their prismatic crystals with a square cross section. Andalusite is commonly found with kyanite, muscovite, corundum, and tourmaline.

Processing Tips: Andalusite can be collected as mineral specimens or used as lapidary material. Chiastolite, a twinned-crystal variety of andalusite, is cut into slabs and used for jewelry. Material can also be tumbled.

ANGLESITE

Category: Sulfate

Crystal System: Orthorhombic

Crystal Habit: Granular, banded, nodular, stalactitic, tabular, prismatic, pseudorhombohedral

Cleavage: Good in one direction

Fracture: Conchoidal

Mohs Scale: 2.5–3

Colors: White, colorless, gray, yellow, green, blue

Luster: Adamantine, vitreous, resinous

Streak: Colorless

Transparency: Transparent, translucent, opaque

Description: Anglesite, a lead sulfate named after Anglesey, Wales, comes in a wide variety of crystal habits and colors. The Romans sought anglesite for its lead content to make many items, particularly water piping.

North American Locations: Mountains in Arizona, Colorado, Idaho, Missouri, New Mexico, and Pennsylvania, and British Columbia, Canada.

Collecting Tips: Anglesite is found in oxidation zones of lead-bearing ore deposits. Due to the nearly two hundred distinct forms of anglesite, it can sometimes be difficult to identify. Much care must be taken when extracting specimens as the crystals can be easily damaged.

Processing Tips: Anglesite is beautiful as both a specimen and polished gemstone. Colorless crystals can be faceted but are difficult to cut, and wearing them in jewelry is not advised as they can be easily damaged.

APATITE

Category: Phosphates
Crystal System: Hexagonal
Crystal Habit: Prismatic, tabular
Cleavage: Poor in one direction
Fracture: Uneven
Mohs Scale: 5

Colors: White, colorless, green, blue, pink, yellow, brown, violet, purple
Luster: Vitreous, subvitreous, resinous, waxy, greasy
Streak: White
Transparency: Translucent, transparent, opaque

Description: The term *apatite* refers to a group of phosphate minerals. *Apatite* is derived from the Greek *apatein*, meaning "to deceive," as it can be easily mistaken for other minerals.

North American Locations: Canada: Mountains in British Columbia and Ontario. United States: Mountains in Alaska, Arizona, California, Idaho, Maine, and Massachusetts, and the hills of northern Florida.

Collecting Tips: Apatite can be found in sedimentary, metamorphic, igneous, and volcanic rocks. Being the "deceiver," apatite can sometimes be difficult to identify by site alone. It is fragile and care must be taken when extracting specimens.

Processing Tips: Fine crystals of apatite look great in any collection, whether in matrix or standalone crystals. Apatite can be cut en cabochon or faceted and is sometimes used in jewelry, although care must be taken as it is a bit soft.

APOPHYLLITE

Category: Phyllosilicate
Crystal System: Tetragonal
Crystal Habit: Prismatic, tabular, massive
Cleavage: Imperfect to perfect in one direction

Fracture: Uneven
Mohs Scale: 4.5–5
Colors: White, colorless
Luster: Vitreous, pearly
Streak: White
Transparency: Transparent, translucent

Description: *Apophyllite* once referred to a specific mineral, but the name is now used for a particular group of phyllosilicates. The mineral derives its name from the Greek *apophylliso*, meaning "to flake off"—a reference to the mineral's propensity to flake apart when exposed to high heat levels.

North American Locations: Canada: Mountains in Quebec. United States: Mountains in Colorado, Idaho, Michigan, New Jersey, and Virginia.

Collecting Tips: Apophyllite is found in vugs in basalt formations and is easily identified by its pyramidal crystal faces. Crystals can be easily broken when extracting, so care must be taken. Wrapping specimens in newspaper, tissue, or Bubble Wrap will help keep them safe when traveling.

Processing Tips: Apophyllite crystals are generally collected as a display mineral. While the mineral can be cut en cabochon or faceted, it is not suitable for jewelry, and should be carefully displayed to avoid any potential damage.

AQUAMARINE

Category: Cyclosilicate
Crystal System: Hexagonal
Crystal Habit: Prismatic to tabular crystals; radial, columnar; granular to compact massive
Cleavage: Fair in one direction

Fracture: Uneven to conchoidal
Mohs Scale: 7.5–8
Colors: Blue, cyan
Luster: Vitreous, resinous
Streak: White
Transparency: Transparent, translucent

Description: Aquamarine is the blue to blue-green gem variety of beryl. It derives its name from the Greek *aqua* and *marina*, meaning "water of the sea." Aquamarine is the birthstone for March. In 1971, Colorado adopted this mineral as the official state gemstone.

North American Locations: Canada: Mountains in British Columbia. United States: Mountains in California, Idaho, Maine, New Hampshire, North Carolina, and especially in the mountains of Colorado.

Collecting Tips: Aquamarine is primarily found in rare beryllium-rich pegmatites and is easily identified by its blue color and crystal shape. Hard-rock tools and chisels must be used to extract specimens, but you will need to take extra care when collecting to avoid damage.

Processing Tips: Fine crystal specimens of aquamarine make for excellent display specimens. It can also be cut en cabochon, faceted, or carved, and used in jewelry.

ARAGONITE

Category: Carbonate
Crystal System: Orthorhombic
Crystal Habit: Pseudohexagonal, prismatic crystals, acicular, columnar, globular, reniform, pisolitic, corralloid, stalactitic, internally banded
Cleavage: Good in one direction
Fracture: Subconchoidal
Mohs Scale: 3.5–4
Colors: White, red, yellow, orange, green, purple, gray, blue, brown
Luster: Vitreous, resinous
Streak: White
Transparency: Transparent, translucent

Description: Aragonite is one of the most common crystal forms of calcium carbonate. It derives its name from Molina de Aragón, Spain, where it was first found. Aragonite is used in saltwater aquariums to replicate reef conditions. It also is useful in cleaning water contaminated by zinc, cobalt, or lead.

North American Locations: Mountains, deserts, plains, and ocean beaches across North America.

Collecting Tips: Aragonite is very fragile. Use care when extracting specimens from host rock sources. It can take many crystal habits, so familiarize yourself with some of them before getting into the field to collect.

Processing Tips: Aragonite crystals only need to be cleaned with soap and room-temperature water and put on display. Massive aragonite is used for ornamental carvings. The mineral is too soft for jewelry, but it can be cut en cabochon and faceted for collectors.

ARFVEDSONITE

Category: Inosilicate, amphibole
Crystal System: Monoclinic
Crystal Habit: Fibrous, radial prismatic aggregates
Cleavage: Perfect in one direction
Fracture: Uneven

Mohs Scale: 5–6
Colors: Black, deep green
Luster: Vitreous
Streak: Deep bluish gray, gray green
Transparency: Translucent, opaque

Description: Arfvedsonite is a type of sodium amphibole and is considered fairly rare. The mineral was named after Swedish chemist Johan August Arfwedson. Arfvedsonite is often misidentified in rock shops as nuummite, a mineral found only in Greenland.

North American Locations: Mountains across North America. Fine specimens come from Arkansas and Colorado, and from Nova Scotia and Quebec, Canada.

Collecting Tips: Arfvedsonite is found in nepheline syenite intrusions and pegmatites and in granites. Look for small dark crystals in vugs of the host rock and talus slopes in areas known for this mineral. Care must be taken extracting specimens.

Processing Tips: Crystals of arfvedsonite tend to be too small for any type of lapidary work, but larger crystals can be faceted. Massive specimens can be tumbled or cut and polished into cabochons, carvings, and display pieces. It is hard enough to use in jewelry, but crystals large enough to facet are exceptionally rare.

AUGITE

Category: Inosilicate
Crystal System: Monoclinic
Crystal Habit: Prismatic, acicular, skeletal, dendritic
Cleavage: Good in two directions
Fracture: Uneven

Mohs Scale: 5.5–6
Colors: Black, brown, greenish
Luster: Vitreous, resinous, dull
Streak: Greenish white
Transparency: Transparent, opaque

Description: A pyroxene, augite is an essential rock-forming mineral. It derives its name from the Greek *augites*, meaning "brightness," in reference to its bright luster. While this mineral is widespread, fine solid collectible crystals are not as common.

North American Locations: Mountains across North America. Fine specimens come from Colorado, New York, and Oregon. The best Canadian specimens come from Ontario.

Collecting Tips: Augite is a ubiquitous rock-forming mineral, so it's almost everywhere as a part of rocks. Fine crystal specimens of augite are not as common but are easily identified by their crystal shape and black color. Some crystals can be brittle, so care must be taken while collecting.

Processing Tips: Fine augite crystals can be cleaned with soap and room-temperature water and put on display. Mineral staining can be removed with light acids. Augite can be cut en cabochon or faceted for collectors and is hard enough for use in jewelry.

AVENTURINE

Category: Silicate
Crystal System: Hexagonal
Crystal Habit: Massive
Cleavage: None
Fracture: Uneven to conchoidal
Mohs Scale: 6.5–7

Colors: Green, red, gold, and brown
Luster: Vitreous, waxy
Streak: White
Transparency: Transparent, opaque

Description: Aventurine is a type of massive quartz that has inclusions of platy minerals that can give the material an optical phenomenon known as aventurescence (a metallic glitter effect). Aventurine was named after a type of glass called aventurine glass. The glass was a chance discovery (*a ventura* is Italian for "by chance").

North American Locations: Mountains, deserts, and river gravels across North America. The best collecting areas are in Oregon and Washington, and British Columbia, Canada.

Collecting Tips: Aventurine comes in many colors but can be identified in the field by its grainy texture. Deposits can be large, and hard-rock tools will be needed to break them down into manageable sizes.

Processing Tips: Aventurine can be tumbled, cut and polished, carved, cut en cabochon, and even faceted. It is commonly used for building materials, headstones, and monuments, as well as in the rough in aquariums and landscaping.

AZURITE

Category: Carbonate
Crystal System: Monoclinic
Crystal Habit: Massive, prismatic, stalactitic, tabular
Cleavage: Good in two directions
Fracture: Conchoidal

Mohs Scale: 3.4–4
Colors: Azure blue, dark to pale blue
Luster: Vitreous
Streak: Light blue
Transparency: Transparent, translucent, opaque

Description: Azurite is a deep blue carbonate hydroxide mineral. Its name is derived from the word *azure*, a French mistranslation of the Persian *lazhward*, which means "blue."

North American Locations: Canada: Mountains and deserts of British Columbia. United States: Mountains and deserts of Arizona, Montana, Nevada, New Mexico, and Utah.

Collecting Tips: Azurite is a common mineral in weathered copper deposits. It is easily identified by its dark blue tone, caused by copper, and is commonly found alongside other copper ores such as malachite, chrysocolla, and cuprite—often combined in the same specimen.

Processing Tips: Azurite is a fairly soft mineral but can still be used in lapidary practices when solid. It can be carved and cut en cabochon, although some care should be used when setting it in jewelry, as it can scratch easily. Fine crystals can be faceted.

BABINGTONITE

Category: Inosilicate
Crystal System: Triclinic
Crystal Habit: Prismatic
Cleavage: Good in two directions
Fracture: Uneven
Mohs Scale: 5.5–6

Colors: Dark green, black
Luster: Vitreous
Streak: Gray
Transparency: Opaque; translucent on thin edges

Description: A rare calcium iron manganese inosilicate, babingtonite was named after Irish physician and mineralogist William Babington. It is not highly sought after by collectors. In 1971, babingtonite was adopted by Massachusetts as its official mineral.

North American Locations: Canada: No known deposits. United States: Northern and central Appalachian Mountains, notably in Connecticut, Massachusetts, New Jersey, and Virginia.

Collecting Tips: Babingtonite is an uncommon mineral that occurs mainly in vugs in various types of igneous rock. It is identified by its dark crystals and prismatic structure and is often found with zeolite minerals, albite, calcite, prehnite, quartz, and garnet. Care should be taken when extracting the fragile crystals.

Processing Tips: Babingtonite is most often cleaned, sometimes trimmed, and used as an in-matrix mineral specimen for display. It can also be faceted, and it's hard enough to be cut en cabochon and used in jewelry.

BARITE

Category: Sulfide
Crystal System: Orthorhombic
Crystal Habit: Bladed, tabular, tubular, fibrous, nodular, massive
Cleavage: Perfect in one direction
Fracture: Uneven

Mohs Scale: 3–3.6
Colors: Colorless, white, yellow, gray, brown, blue
Luster: Vitreous, pearly
Streak: White
Transparency: Transparent, opaque

Description: Also spelled *baryte* or *barytes*, barite is barium sulfide. Its name is derived from the Greek *baryos*, meaning "heavy," as barite has a very high specific gravity. It is used industrially as a source of barium used in the making of paper and rubber.

North American Locations: Canada: Mountains and plains of British Columbia. United States: California, Colorado, Nevada, and South Dakota.

Collecting Tips: Barite is fairly common across North America in both igneous and sedimentary environments. The heavy weight of the crystals is a good identifying indicator of barite. It's a brittle and soft mineral, so care must be taken when digging and transporting specimens.

Processing Tips: Barite is mainly collected as a mineral specimen. Stained crystals can be soaked in light acid solutions to remove deposited minerals. Care must be taken when processing barite. It can be faceted but is too soft for jewelry.

BAUXITE

Rock Type: Sedimentary

Structure: Massive, earthy, micro-crystalline, often with small spherical bodies

Major Minerals: Aluminum oxides and hydroxides

Minor Minerals: Hematite, various clay minerals

Fossils: None

Colors: Gray, yellow, reddish, brown

Texture: Coarse-grained

Description: Bauxite is a type of sedimentary rock with a high aluminum content and is mined for the aluminum to make everyday products. It is the only commercial ore of aluminum. Its name is derived from the name of the village it was discovered near: Les Baux-de-Provence, France.

North American Locations: Canada: No known deposits. United States: Hills and plains of Alabama and Arkansas; small deposits also exist in the deserts of California and in the hills of Georgia and Virginia.

Collecting Tips: Most bauxite locations in the United States are claimed and mined for the aluminum. The savvy rockhound may be able to gain permission from mine operators to collect a few specimens. Some rock clubs also make field trips to bauxite mines.

Processing Tips: Unless you plan on digging several tons of bauxite to process into aluminum, there's not a lot that can be done with it besides using a piece as a specimen for your collection. It can be cut and polished.

BELEMNITE

Category: Fossil
Mohs Scale: 3–4
Colors: Gray, tan, brown, brownish black

Luster: Earthy, dull
Transparency: Opaque

Description: Belemnites are an extinct type of cephalopod that lived from the Jurassic to Late Cretaceous periods. They resembled modern-day squids but had a calcitic guard, or shell, that protected the animal from predators. The calcitic guard is the part of the belemnite that is found today as a fossil. Belemnite was adopted as the state fossil of Delaware in 1996.

North American Locations: Mountains, deserts, plains, and lake and ocean beaches across North America.

Collecting Tips: Belemnites are found in marine sedimentary rocks and can be easily identified by their bullet-shaped calcitic guard, or shell. They can also display suture lines common in cephalopod fossils. Care must be taken when extracting belemnites from their hard host rock as they can be easily broken. Broken specimens can be glued back together.

Processing Tips: Fossils can be put in a display as found in nature if the shells are exposed well enough. An experienced prepper can expose more of the fossil with specialty tools such as an air scribe to give specimens a more aesthetically pleasing look.

BENITOITE

Category: Cyclosilicate
Crystal System: Hexagonal
Crystal Habit: Tabular to dipyramidal crystals, granular
Cleavage: None
Fracture: Uneven to conchoidal

Mohs Scale: 6–6.5
Colors: Blue, colorless
Luster: Vitreous
Streak: White
Transparency: Translucent, transparent

Description: A rare barium titanium cyclosilicate, benitoite was discovered at San Benito Mountain in Southern California in 1907. It was first thought to be sapphire due to its deep blue color. California is the only location to produce gem-quality specimens.

North American Locations: Canada: No known deposits. United States: The most important benitoite deposit is in the mountains of Southern California; small amounts of benitoite are occasionally found in the mountains of Arkansas and Montana.

Collecting Tips: As a rare and valuable mineral, the few known locations are claimed and privately mined. Sometimes local rock clubs may be allowed into the mines to dig through tailings. The mines will also sometimes sell mail-order bags of unprocessed gem-bearing ore.

Processing Tips: Benitoite is often collected as a crystal specimen. This rare mineral has a higher light dispersion than diamond and makes for a wonderful faceted gem. Lower-grade specimens can be cut en cabochon. It is fairly hard and can be used in jewelry.

BIOTITE

Category: Phyllosilicate
Crystal System: Monoclinic
Crystal Habit: Platy, massive
Cleavage: Perfect in one direction
Fracture: Uneven
Mohs Scale: 2.5–3

Colors: Dark brown, greenish brown, brownish black, yellow, white
Luster: Vitreous, pearly
Streak: White
Transparency: Translucent, transparent, opaque

Description: Named after French physicist Jean-Baptiste Biot, biotite refers to a family of phyllosilicate minerals in the mica group. Masses of platy crystals of mica are called books because they look like the pages of a thick book. Biotite helps scientists calculate minimum ages of rocks.

North American Locations: All geological settings across North America.

Collecting Tips: Biotite is abundant and can be found in igneous rock. It can be easily recognized by the dark masses of platy crystals.

Both biotite and muscovite mica are good indicators of pegmatite veins, which also contain gems such as beryl, topaz, and tourmaline.

Processing Tips: It is not recommended to use water or liquid cleaners on biotite as they can soak into the mineral and cause it to eventually break apart. A dry electric toothbrush is the best tool for cleaning mica. Biotite doesn't cut or polish well; specimens are mostly used for display.

CALCITE

Category: Carbonate
Crystal System: Trigonal
Crystal Habit: Crystalline, granular, stalactitic, concretionary, massive, rhombohedral

Cleavage: Perfect in three directions
Fracture: Conchoidal, but difficult to discern
Mohs Scale: 3

Colors: Colorless, white, yellow, red, orange, blue, green, brown, gray
Luster: Vitreous to pearly
Streak: White
Transparency: Transparent, translucent

Description: Calcite is one of the three most common carbonate minerals found in the earth's crust, along with dolomite and siderite. Colorless calcite crystals exhibit the optical property known as double refraction, meaning they split an image into two reflections.

North American Locations: All geological settings across North America.

Collecting Tips: Calcite is a component of many rocks, minerals, and fossils found throughout North America. It is a soft and fragile mineral, so extra care should be taken when extracting it. Massive calcite, on the other hand, can be cleaved into rhombohedral crystals with less care.

Processing Tips: Calcite is a sensitive mineral to process. Even light acid baths will completely dissolve calcite and calcium-based minerals. With care, calcite can be tumbled, carved, cut, and polished—even cut en cabochon and faceted. It is not recommended for jewelry.

CARNELIAN

Category: Silicate
Crystal System: Hexagonal
Crystal Habit: Microcrystalline
Cleavage: None
Fracture: Uneven to conchoidal
Mohs Scale: 6–7

Colors: Red, orange, brownish red
Luster: Vitreous, dull, greasy, silky
Streak: White
Transparency: Translucent

Description: Carnelian is a type of chalcedony with orange to deep red tones due to traces of iron. It has been used since Neolithic times for a wide range of products, from ornamental beads to arrowheads. Today carnelian is still a highly sought-after mineral for lapidary and knapping. Its name comes from the Latin *carneus*, meaning "fleshy."

North American Locations: Deserts, river gravels, and lake and ocean beaches across North America. Fine specimens come from the deserts of Arizona, California, Nevada, New Mexico, and Utah.

Collecting Tips: Carnelian is easily recognized by its bright red and orange tones and vitreous to greasy luster. It will often show conchoidal fractures, which look like a finger was pressed into the stone or like the edge of broken glass.

Processing Tips: Carnelian is an excellent lapidary material that takes a great polish and glows when backlit. It is a popular tumbling material, used for carvings and decorations, and superb for cutting cabochons for jewelry.

CARNOTITE

Category: Vanadate
Crystal System: Monoclinic
Crystal Habit: Earthy, powdery
Cleavage: Perfect in one direction
Fracture: Uneven
Mohs Scale: 2

Colors: Bright yellow to greenish yellow
Luster: Dull
Streak: Yellow
Transparency: Opaque

Description: Carnotite is a potassium uranyl vanadate and an important ore of uranium. It is named for French chemist Marie-Adolphe Carnot. It is mildly radioactive.

North American Locations: Canada: Hills of Alberta, Canada. United States: Deserts of Arizona, Colorado, New Mexico, and Utah, and the hills of eastern Pennsylvania.

Collecting Tips: Carnotite occurs as coatings and crusts on sandstone and is easily identified by its bright yellow color. Specimens can be collected with rock picks, but care must be taken not to damage the delicate coatings.

Processing Tips: Carnotite specimens should be cleaned only with water. Avoid contact with any hard objects.

CASSITERITE

Category: Oxide
Crystal System: Tetragonal
Crystal Habit: Pyramidal, prismatic, radially fibrous, botryoidal crusts, and coarse- to fine-grained concretionary masses

Cleavage: Perfect in one direction; fair in another
Fracture: Uneven
Mohs Scale: 6–7
Colors: Black, brownish black, reddish brown, brown, red, yellow, gray, white

Luster: Adamantine to submetallic
Streak: White to brownish
Transparency: Translucent, transparent, opaque

Description: Cassiterite is a tin oxide and the most important ore of the metal. Its name is derived from the Greek *kassiteros*, meaning "tin."

North American Locations: Canada: Mountains in Ontario and Yukon Territory. United States: Mountains in California, Colorado, and North Carolina.

Collecting Tips: Crystals of cassiterite are easily identified in the field by their crystal habit and heavier-than-average weight. Most found crystals will have survived rolling down waterways, but the crystal shape may or may not be recognizable. Care should be used removing rare in-matrix specimens.

Processing Tips: Transparent cassiterite crystals are an excellent material for faceting, but they are not commonly seen in the mainstream marketplace, as most material mined is used for industry. It can also be carved, tumbled, and cut en cabochon. Crystals found in matrix can be trimmed, cleaned, and put on display.

CAVANSITE

Category: Silicate
Crystal System: Orthorhombic
Crystal Habit: Radiating acicular prismatic crystals commonly as spherulitic rosettes
Cleavage: Good in one direction

Fracture: Uneven to conchoidal
Mohs Scale: 3–4
Colors: Sky blue to greenish blue
Luster: Vitreous, pearly
Streak: Bluish white
Transparency: Transparent

Description: Cavansite was originally discovered in Oregon in 1967 when a dam was being built on the Owyhee River. This is the only known location to find specimens in North America. Its name is derived from its chemical composition: **ca**lcium **van**adium **sili**ca**te**.

North American Locations: Canada: No known deposits. United States: The high desert of Malheur County, Oregon.

Collecting Tips: Should you find yourself near the Owyhee Dam, cavansite can be found in vugs in a brown tuff that is partially filling a fault fissure. This known exposure was covered by the construction of the dam, but tenacious rockhounds may be able to find material nearby.

Processing Tips: When found, cavansite is kept in its natural state. The matrix of specimens may be trimmed for better display purposes, but care must be taken when doing so, as it can be easily broken.

CELESTINE

Category: Sulfate
Crystal System: Orthorhombic
Crystal Habit: Tabular to pyramidal
Cleavage: Perfect in one direction
Fracture: Uneven
Mohs Scale: 3–3.5

Colors: White, pink, pale green, pale brown, black, pale blue, reddish, gray
Luster: Vitreous, pearly on cleavages
Streak: White
Transparency: Transparent, translucent

Description: Also called celestite, celestine is a strontium sulfate known for its beautiful blue tones. Strontium is the element used to create the red in fireworks and flares and is also used in the creation of metal alloys.

North American Locations: Plains, dry-lake beds, lake shorelines, and evaporite deposits in Ontario and Quebec. United States: Plains, dry-lake beds, lake shorelines, and evaporite deposits in Arizona, California, Michigan, New York, and Ohio.

Collecting Tips: Celestine can be found mainly in evaporite deposits in sedimentary outcrops, often with/ near halite (salt). The crystals can be quite fragile, and care must be taken when extracting them from the host rock. Chisels and hammers will be needed to remove crystal clusters. Bubble Wrap is recommended for transporting the mineral.

Processing Tips: Dry brushing with a soft toothbrush is the best way to clean celestine. While water is mostly okay for cleaning, it can sometimes cause crystals to become cloudy. Larger broken crystals can be hand-shaped, polished, and faceted, but it is difficult to cut, so cutting is only recommended for experienced lapidaries.

CERUSSITE

Category: Carbonate
Crystal System: Orthorhombic
Crystal Habit: Massive granular, reticulate, tabular to equant
Cleavage: Good in one direction
Fracture: Conchoidal
Mohs Scale: 3–3.5

Colors: Colorless, white, gray, blue, green
Luster: Adamantine, vitreous, resinous
Streak: White
Transparency: Transparent, translucent

Description: Cerussite is a lead carbonate and one of the more important ores of lead. The name is derived from the Latin *cerussa*, meaning "white lead." In the past it was used to make lead paint and cosmetics.

North American Locations: Mountains and deserts of western North America.

Collecting Tips: Cerussite is found in the oxidized zones of base metal deposits, particularly in lead-silver deposits. Special care should be taken when extracting crystals, not only because they are fragile, but also because they contain lead. Gloves should be worn when mining this mineral.

Processing Tips: Crystals of cerussite are often cleaned, prepped, and put on display. It is possible to facet this material, but it is difficult to cut without breaking. It is not recommended for use in jewelry, as it is too soft.

CHABAZITE

Category: Tectosilicate
Crystal System: Trigonal
Crystal Habit: Pseudocubic, rhombohedral
Cleavage: Poor in three directions
Fracture: Irregular

Mohs Scale: 4–5
Colors: Colorless, white, yellow, pink, orange, brown
Luster: Vitreous
Streak: White
Transparency: Transparent, translucent

Description: Chabazite is a member of the large zeolite family and is similar in structure and habit to its close relative, gmelinite. It derives its name from the Greek *chabazios*, meaning "hailstone." The word originated in Orpheus's "Lithica," an ancient poem about the magical properties of minerals.

North American Locations: Mountains across North America. Fine specimens come from New Jersey and Colorado.

Collecting Tips: Chabazite is most commonly found in the cavities, veins, and vugs in basalt. It can be identified in the field by its crystal habit and brown-orange color tones. Care should be taken when extracting crystal clusters from the host rock so as not to damage them. Chisels and hammers are helpful.

Processing Tips: Fine colorless crystals can be faceted but are too soft for jewelry. Crystals can usually be cleaned with light soap, water, and a soft toothbrush. The matrix can be trimmed for a more aesthetically pleasing display.

CHALCEDONY

Category: Oxide
Crystal System: Hexagonal
Crystal Habit: Microcrystalline
Cleavage: None
Fracture: Conchoidal
Mohs Scale: 6–7

Colors: Often colorless, but can be any color
Luster: Waxy, vitreous, dull, greasy, silky
Streak: White
Transparency: Translucent, transparent

Description: Chalcedony is often referred to as agate, and while they are the same mineral, agate is chalcedony displaying some sort of pattern or inclusion, such as banding, moss, or plumes. Chalcedony derives its name from the ancient port Khalkedon (in modern-day Turkey), where deposits of the mineral were found.

North American Locations: Deserts, mountains, and river gravels, and along ocean and lake shorelines, across North America.

Collecting Tips: Chalcedony can be recognized in the field by its translucency, uneven fracture, and waxy to dull luster. It can be just about any color, but it is most commonly found colorless or in light yellow or gray tones.

Processing Tips: Chalcedony is a popular lapidary material for carvings, cabochons, faceting, and even stained glass windows. It takes an excellent polish and can be used in all jewelry applications. Agates are the most popular mineral used in tumblers.

CHALCOPYRITE

Category: Sulfide
Crystal System: Tetragonal
Crystal Habit: Sphenoidal resembling tetrahedrons; aggregate; massive
Cleavage: Poor in one direction
Fracture: Uneven

Mohs Scale: 3.5–4
Colors: Brass yellow, oxidized purples and blues
Luster: Metallic
Streak: Greenish black
Transparency: Opaque

Description: Chalcopyrite is a copper sulfide, and it is the most commonly mined ore of copper. It derives the first part of its name from the Greek *khalkos*, which means "copper." The second part of its name, *pyrite*, is a reference to the mineral's chemical composition.

North American Locations: Mountains and deserts across western North America, notably in Arizona, New Mexico, Utah, Colorado, and Montana, and in British Columbia, Canada.

Collecting Tips: Chalcopyrite is recognized in the field by its bright, brassy coloration, especially when freshly exposed. Weathered material can have an iridescent, tarnished surface. Chalcopyrite is a very heavy mineral. Hard-rock mining tools will be required for extracting material.

Processing Tips: Large masses of chalcopyrite can be carved, cut, and polished, but the material tends to be brittle and is not recommended for the beginner lapidary. Most is kept as an ore specimen. Some rockhounds soak it in a light acid bath to create a more iridescent surface.

CHERT

Rock Type: Sedimentary
Structure: Massive, sometimes bedded
Major Minerals: Quartz
Minor Minerals: Hematite

Fossils: None
Colors: Gray, brownish, reddish
Texture: Smooth, very fine-grained

Description: This hard, durable rock consists mainly of quartz and occurs in a wide range of colors. With its fine grain and distinct conchoidal fracture, it was knapped into tools and weapons by many Stone Age cultures.

North American Locations: Deserts, forests, mountains, plains, and river gravels, and along lake and ocean shorelines across North America.

Collecting Tips: Because of its durability, chert often appears as larger pieces of rock amid sand and gravel surfaces. Larger rocks on the floor of sandy deserts are likely to be chert.

Processing Tips: Chert is easily cleaned with water and a stiff brush. It is a fine rock for tumbling.

CHLORASTROLITE

Category: Sorosilicate
Crystal System: Monoclinic
Crystal Habit: Fibrous aggregates
Cleavage: None
Fracture: Subconchoidal
Mohs Scale: 5–6

Colors: Green, bluish green, white
Luster: Silky, pearly
Streak: White
Transparency: Opaque

Description: Also called Isle Royale greenstone, chlorastrolite is a variety of the mineral pumpellyite. It derives its name from the Greek *khloritis*, meaning "green," and *asterismos*, meaning "marking with stars." In 1973, chlorastrolite was adopted as the state gem of Michigan.

North American Locations: Almost exclusively in the forests of Michigan's Keweenaw Peninsula.

Collecting Tips: Chlorastrolite has a distinct green and white radiating turtleback pattern and is highly chatoyant, making it easy to spot in the field.

Processing Tips: Great care must be taken when processing chlorastrolite. The fine-webbed pattern can easily be ground away by the inexperienced lapidary when making cabochons or tumbling. The mineral is somewhat hard and takes an excellent polish on the lapidary wheel.

CHRYSOCOLLA

Category: Phyllosilicate
Crystal System: Orthorhombic
Crystal Habit: Massive
Cleavage: None
Fracture: Uneven to subconchoidal
Mohs Scale: 2–3

Colors: Blue, blue green
Luster: Vitreous, earthy
Streak: Pale blue, tan, gray
Transparency: Translucent, opaque

Description: Chrysocolla is formed by the decomposition of copper minerals. It derives its name from the Greek *chrysos*, which means "gold," and *kolla*, which means "glue." In ancient Greece, the term *chrosocolla* referred to many different materials in soldering gold.

North American Locations: Arid regions and deserts of western North America, notably in Arizona, California, Colorado, Idaho, Nevada, New Mexico, Montana, and Texas, and in British Columbia, Canada. Specimens are occasionally found in the hills of Connecticut and Virginia.

Collecting Tips: Chrysocolla can be found near copper deposits. The striking blue color of this mineral, contrasted against drab desert host rock, is easily spotted in the field. It is commonly found with azurite, cuprite, and malachite.

Processing Tips: Chrysocolla can be very soft, so care must be taken when polishing and grinding this material. Always wear respirators when working with stone, but especially copper minerals, as they can be harmful to breathe. Chrysocolla can be used to make carvings, beads, and cabochons.

CINNABAR

Category: Sulfide
Crystal System: Trigonal
Crystal Habit: Rhombohedral and tabular
Cleavage: Perfect in one direction
Fracture: Uneven

Mohs Scale: 2–2.5
Color: Cochineal red
Luster: Adamantine, dull
Streak: Scarlet red
Transparency: Transparent, opaque

Description: Cinnabar is a type of mercury sulfide and is one of the largest resources of the highly toxic metal. It was also once crushed into a powder and used as paint pigment, though it has since been replaced by less toxic synthetic varieties. Its name is derived from the Arabic *zinjafr* and also the Persian *zinjirfrah*, both meaning "dragon's blood."

North American Locations: Mountains in western North America, notably in California, Oregon, Nevada, and Texas, and in British Columbia, Canada.

Collecting Tips: Cinnabar is recognized in the field, usually in areas of hot springs and volcanic rocks, by its bright cochineal red coloring. Extreme caution must be taken when collecting this toxic mineral. Face masks and gloves should be used. Don't touch your eyes or mouth during collecting, and wash well after collecting.

Processing Tips: If collected, cinnabar should be put on display and rarely touched. It's too soft to work or take a polish unless it's an inclusion in other minerals such as common opal or chalcedony.

CITRINE

Category: Tectosilicate
Crystal System: Hexagonal
Crystal Habit: Prismatic
Cleavage: None
Fracture: Conchoidal
Mohs Scale: 7

Colors: Yellow to yellow brown
Luster: Vitreous
Streak: Colorless
Transparency: Translucent, transparent, opaque

Description: Citrine is the yellow or yellow-brown variety of crystalline quartz. Natural citrine can be very rare to find in nature. Much of what is sold on the world market as citrine is actually heat-treated amethyst. It derives its name from the Latin *citrina*, meaning "citrus."

North American Locations: Canada: Mountains of British Columbia. United States: Mountains in California, Colorado, New Hampshire, and North Carolina, and of British Columbia, Canada.

Collecting Tips: Citrine is recognized in the field by its bright to smoky yellow coloring. Sometimes it can mix with smoky quartz and be a bit more difficult to identify. Ametrine, a combination of citrine and amethyst, is an even rarer occurrence.

Processing Tips: The lapidary can carve or cut citrine en cabochon, but faceted citrine reveals all its beauty. Cabochons and faceted citrine make for excellent jewelry material. Fine crystals can also be left as found in the field.

| Rock | **COAL** |

Rock Type: Sedimentary, altered
Structure: Variable: porous to solid, splintery to fibrous
Major Minerals: Carbon
Minor Minerals: Sulfur, quartz, hematite

Fossils: Often contains plant-fossil impressions
Colors: Jet black to yellowish brown
Texture: Usually fine-grained

Description: There are many varieties of coal, but what most people consider coal is anthracite, which is dug for heating and coal-burning power plants. Anthracite contains the highest percentages of carbon of the different coal family members.

North American Locations: Mountains, deserts, forests, and plains of North America. The largest anthracite deposit in the world is found in Lackawanna County, Pennsylvania.

Collecting Tips: Be sure to wear gloves, as well as clothes you don't mind getting dirty, as coal is one of the messiest rocks a rockhound can collect. It can be easily identified by its black, almost metallic, luster. It will also be much lighter in weight compared to most other rocks.

Processing Tips: Besides carving, anthracite isn't much good for most lapidary projects. Most pieces collected are simply put on display.

COPPER

Category: Native element
Crystal System: Cubic
Crystal Habit: Massive
Cleavage: None
Fracture: Hackly
Mohs Scale: 2.5–3

Colors: Copper red, brown
Luster: Metallic
Streak: Rose
Transparency: Opaque

Description: Native copper was likely the first metal used by humans. It derives its name from the Latin *aes Cyprium*, which means "metal of Cyprus." Copper is used industrially for electrical wiring, building materials, and in industrial machinery.

North American Locations: Canada: Eastern end of Lake Superior in Ontario. United States: Mountains and deserts of Arizona, New Mexico, and Utah, but mainly in the forests of the Keweenaw Peninsula of Michigan.

Collecting Tips: Native copper found in the field will be highly oxidized and will not look like a shiny new penny. However, copper can be easily identified by its very heavy weight. Some rockhounds use metal detectors to locate specimens.

Processing Tips: Oxidized native copper specimens can be immersed in muriatic acid to remove the patina before being cleaned, dried, and then polished with common store-bought copper polish. Specimens are commonly ground and polished on one side and left with a natural patina on the other for contrast.

COVELLITE

Category: Sulfide
Crystal System: Hexagonal
Crystal Habit: Foliated
Cleavage: Perfect in one direction
Fracture: Uneven
Mohs Scale: 1.5–2

Colors: Indigo blue, black
Luster: Submetallic, resinous
Streak: Lead gray, black
Transparency: Opaque

Description: Covellite is a somewhat rare copper sulfide. It was the first naturally occurring superconductor and is used industrially for a wide variety of purposes. It was named after Niccolo Covelli, who was the first to describe the mineral in writing.

North American Locations: Canada: Mountains and deserts of British Columbia and Ontario. United States: Mountains and deserts of Alaska, Colorado, Montana, Utah, and Wyoming.

Collecting Tips: Covellite is found as vein and replacement deposits in copper-rich areas. It is recognized in the field by its iridescent, metallic blue tones and foliated tabular crystals. It is also quite heavy when compared to other common minerals. It is soft, so care should be taken when extracting the mineral from host rock.

Processing Tips: Covellite can be kept as an ore specimen and looks quite striking on its own, but it can also be polished. The mineral is too soft for use in most jewelry, and a light touch should be used when grinding and polishing any specimen.

CRINOID

Category: Fossil
Mohs Scale: 3–4
Colors: Gray, tan, brown
Luster: Earthy, dull

Transparency: Opaque; thin sections may be translucent

Description: Crinoids are marine animals of the echinoderm group, which includes starfish and sea urchins. They consist of a segmented stem, branches, and leaflike arms. Crinoids frequently become fossilized in limestone and shale.

North American Locations: Mountains, plains, forests, river gravels, and lake and ocean shorelines across North America.

Collecting Tips: The best crinoid-collecting sites are exposures of weathered limestone and shale. Screens are helpful in recovering small fossils such as the disklike segments of crinoid stems.

Processing Tips: Crinoid fossils are fragile and should be cleaned only with water and a soft brush.

CUPRITE

Category: Oxide
Crystal System: Cubic
Crystal Habit: Octahedral
Cleavage: Fair in three directions
Fracture: Conchoidal
Mohs Scale: 3.5–4

Colors: Red, black, gray
Luster: Adamantine, submetallic
Streak: Brownish red
Transparency: Transparent, opaque

Description: One of the most important ores of copper, cuprite is a secondary mineral that is formed by oxidizing copper sulfide veins. It derives its name from the Latin *cuprum*, meaning "copper," and is recognized by its red coloration.

North American Locations: Mountains and deserts of western North America, notably in Arizona, Montana, Nevada, New Mexico, and Utah, and British Columbia, Canada.

Collecting Tips: Cuprite is recognized in the field by its octahedral crystal habit and brownish red color, although it can sometimes have a darker, more gray or black coloration. Care should be taken when extracting crystals from hard host rock so as not to damage them.

Processing Tips: Cuprite crystals in matrix make for excellent display specimens. Sometimes the host rock will need to be trimmed so the specimen can sit better in a display. Large transparent crystals can be faceted for collectors. It is too soft for most jewelry applications.

DANBURITE

Category: Sorosilicate
Crystal System: Orthorhombic
Crystal Habit: Prismatic
Cleavage: Poor in one direction
Fracture: Uneven to subconchoidal
Mohs Scale: 7–7.5

Colors: Colorless, yellow
Luster: Vitreous, greasy
Streak: White
Transparency: Transparent, translucent

Description: Danburite is chemically a calcium borosilicate that derives its name from the city of Danbury, Connecticut, where the mineral was first discovered. It is mainly mined as a gemstone for the jewelry and metaphysical markets.

North American Locations: Canada: No known deposits. United States: Forests and hills of Connecticut and New York.

Collecting Tips: Danburite can be found in granite, metamorphosed carbonate, and evaporite deposits. It is often mistaken for topaz but is easily distinguished by its lower brilliance and poor cleavage. Due to its poor cleavage, care should be taken when extracting and transporting specimens in the field, as they can break easily.

Processing Tips: Danburite crystals can be simply cleaned, prepped, and put into a collection as found. While fragile, it is also a popular material with experienced lapidaries for cutting en cabochon, carving, and faceting.

DIAMOND

Category: Native element
Crystal System: Cubic
Crystal Habit: Octahedral, cubic
Cleavage: Perfect in four directions
Fracture: Conchoidal
Mohs Scale: 10

Colors: White, black, colorless, yellow, pink, red, blue, brown
Luster: Adamantine
Streak: None (will scratch streak plate)
Transparency: Transparent, translucent, opaque

Description: Diamond is pure crystalline carbon and is the hardest known natural mineral formed on earth. It derives its name from the Greek *adamas*, meaning "invincible," a reference to the mineral's hardness. High-quality diamonds are very popular gemstones for jewelry, and low-quality diamonds are used as an industrial abrasive.

North American Locations: Concentrations of diamonds are found only across the tundra of northern Canada; the mountains of Colorado; and the hills of Arkansas, which is the site of Crater of Diamonds State Park.

Collecting Tips: Diamond is a very heavy mineral and likes to settle into nooks and crannies, much like gold. There are many stories of gold hunters finding diamonds while panning. Rough diamonds are also very slick and don't stay pinched between fingers for long.

Processing Tips: Diamonds are a popular material for faceting. High-grade diamonds are sent to professional cutters, while low-grade crystals are better left as specimens on display.

DIOPSIDE

Category: Inosilicate
Crystal System: Monoclinic
Crystal Habit: Equant, prismatic
Cleavage: Good in two directions
Fracture: Uneven
Mohs Scale: 6

Colors: White, green, violet blue
Luster: Vitreous
Streak: White, pale green
Transparency: Transparent, translucent

Description: Diopside is a rock-forming member of the pyroxene family of minerals and derives its name from the Greek *di*, meaning "two," and *opsis*, meaning "vision." This is a reference to the two ways it orients the vertical prism. Bright green diopside colored by chromium is referred to as chrome diopside. The blue variety, colored by manganese, is often referred to as violane.

North American Locations: Canada: Forests and mountains in Ontario. United States: Forests and mountains in California and New York.

Collecting Tips: Diopside is found in olivine-rich basalts and andesites, kimberlite, and crystalline limestones and dolomites. Chrome diopside is identified in the field by its bright green or blue tones. Colorless crystals can be identified by crystal structure and a scratch test.

Processing Tips: Diopside is beautiful as a rough specimen but can also be used for lapidary and jewelry purposes. Fine transparent crystals are often faceted, and lower-grade material is cut en cabochon.

DIOPTASE

Category: Cyclosilicate
Crystal System: Hexagonal
Crystal Habit: Prismatic
Cleavage: Perfect in three directions
Fracture: Uneven to conchoidal

Mohs Scale: 5
Colors: Emerald, blue green
Luster: Vitreous, greasy
Streak: Pale bluish green
Transparency: Transparent, translucent

Description: Dioptase is a copper cyclosilicate that derives its name from the Greek *dia* ("through") and *optos* ("visible"). This is a reference to an observer being able to see the mineral's internal cleavage planes. It is those same perfect planes that make dioptase a poor mineral to cut for jewelry.

North American Locations: Canada: No known deposits. United States: Deserts of Arizona, Nevada, and New Mexico.

Collecting Tips: Dioptase can be found where oxidized zones of copper sulfide mineral deposits are located. It is identified by its dark blue-green color, similar to emerald, and its prismatic crystal habit. Extra care must be taken when extracting dioptase, as it will break easily with a wrong swing of a hammer or chisel.

Processing Tips: Dioptase is a fragile stone and not often used for lapidary purposes. Specimens can be cleaned, possibly prepped, and added as a specimen of rare crystal in a collection.

DOLOMITE

Category: Carbonate
Crystal System: Trigonal
Crystal Habit: Rhombohedral, prismatic, botryoidal, massive, grainy
Cleavage: Perfect in three directions
Fracture: Uneven to conchoidal

Mohs Scale: 3.5–4
Colors: Colorless, white, cream
Luster: Vitreous
Streak: White
Transparency: Transparent, translucent

Description: Dolomite is a type of calcium magnesium carbonate common throughout North America. The mineral was named after French mineralogist Déodat de Gratet de Dolomieu. Dolomite is a minor ore of magnesium and mined for use in the medical industry, as well as in construction, gardening, and metal alloys.

North American Locations: Forests, hills, and plains of central North America, notably in Missouri and Tennessee, and in Canada's Ontario Peninsula between lakes Erie, Ontario, and Huron.

Collecting Tips: Dolomite occurs in formations of limestone and dolomite rock. Care should be taken when collecting these crystals, as they can easily break when being extracted from the host rock. When collecting for carving, find the largest pieces you can in talus, or use hard-rock mining tools to extract large pieces from deposits.

Processing Tips: Massive dolomite can be easily carved and shaped. Fine crystals of dolomite can be cleaned with light soap and water. Do not use acid when cleaning dolomite, as it can easily etch and even dissolve the mineral.

Mineral

EMERALD

Category: Cyclosilicate
Crystal System: Hexagonal
Crystal Habit: Prismatic
Cleavage: Fair in one direction
Fracture: Uneven to conchoidal
Mohs Scale: 7.5–8

Color: Green
Luster: Vitreous
Streak: White
Transparency: Transparent, translucent

Description: Emerald is the green-toned variety of the beryl family. It owes its unique green hue to traces of chromium, vanadium, and iron. Its name has derivations from many languages, each translating to "green stone." Emerald is the birthstone for May.

North American Locations: Canada: Forests and mountains in Yukon Territory. United States: Forests and mountains in North Carolina.

Collecting Tips: Emerald is found in granitic pegmatites and areas of contact metamorphism and is recognized in the field by its bright green color and hexagonal crystals. While a strong and hard mineral, crystals should be extracted from host rocks carefully.

Processing Tips: Emerald makes for beautiful, faceted gemstones to be set in jewelry. It can also be cut en cabochon, carved, tumbled, or put on display as is. Crystals with any stubborn staining can be cleaned with a weak acid solution.

EPIDOTE

Category: Sorosilicate
Crystal System: Monoclinic
Crystal Habit: Prismatic
Cleavage: Perfect in one direction
Fracture: Uneven
Mohs Scale: 6–7

Color: Green
Luster: Vitreous
Streak: Colorless, grayish
Transparency: Translucent, transparent, opaque

Description: Epidote is a calcium aluminum iron silicate. It derives its name from the Greek *epidosis*, meaning "donation" or "addition," a reference to the mineral having one side of the prism longer than the other.

North American Locations: Epidote is widespread across North America; excellent collecting sites are located in the forests, mountains, and river gravels of Alaska, California, Colorado, and Idaho, and of Ontario, Canada.

Collecting Tips: Epidote crystals are identified in the field by their dark green tones and crystal habit. Massive epidote weathered by rivers and tides can be a bit more difficult to identify due to the large number of different types of green rocks in the field.

Processing Tips: Crystals can be cleaned with room-temperature water. For more stubborn staining, a light acid solution can be used. Massive specimens of epidote mixed with other minerals, as well as broken crystals, can be carved, tumbled, or cut en cabochon. Transparent crystal specimens can be faceted.

EUDIALYTE

Category: Cyclosilicate

Crystal System: Trigonal

Crystal Habit: Rhombohedral, prismatic, granular

Cleavage: Perfect in one direction

Fracture: Uneven

Mohs Scale: 5–6

Colors: Red tones, brown, blue, yellow

Luster: Vitreous

Streak: White

Transparency: Transparent, translucent

Description: Eudialyte is a fairly rare cyclosilicate mineral. It derives its name from the Greek *eu dialytos*, which means "readily decomposable," a reference to the mineral's solubility in acids. Eudialyte is a minor ore of zirconium but is mainly used as a gemstone in jewelry and collecting.

North American Locations: Canada: Hills at Mont-Saint-Hilaire, Quebec. United States: Forests and mountains in Alaska, Arkansas, and New Mexico.

Collecting Tips: Eudialyte forms in alkaline igneous rocks. Fine crystals are recognized by their bright red tones and crystal habit. Extreme care should be taken when extracting eudialyte, as the crystals break easily. Use paper or plastic wrapping to protect crystals during transport.

Processing Tips: Extremely rare and fragile, eudialyte should only be cut and polished by seasoned lapidaries. Cleaning, possibly trimming the matrix, and putting it on display is common. Do not clean eudialyte with acids.

FLUORITE

Category: Halide
Crystal System: Cubic
Crystal Habit: Cubic, octahedral
Cleavage: Perfect in four directions
Fracture: Uneven
Mohs Scale: 4

Colors: Any
Luster: Vitreous
Streak: White
Transparency: Transparent, translucent

Description: Fluorite is a calcium fluoride mineral (hence its name). It is highly collectible due to its wide range of colors and collecting sites. Industrially, fluorite is mined for use in manufacturing hydrofluoric acid, steel, and high-octane fuels.

North American Locations: Canada: Hills, mountains, and plains of Ontario. United States: Colorado, Illinois, Kentucky, New Mexico, and Ohio.

Collecting Tips: Fluorite is most commonly found in felsic igneous rocks, granitic pegmatites, and limestones. It can be recognized in the field by its crystal habit. It comes in every color, although common tones are green, purple, blue, and colorless. It is soft and can cleave easily, so care must be taken when extracting specimens.

Processing Tips: Care must be taken when processing this soft mineral. Heavily stained crystals can be cleaned with a light mixture of muriatic acid. Massive fluorite can be carved and turned into cabochons, although it is not recommended for use in most jewelry due to its softness and perfect cleavage.

FULGURITE

Category: Silicate
Crystal System: None, amorphous
Crystal Habit: None
Cleavage: Varies
Fracture: Varies
Mohs Scale: Varies

Colors: Brown, tan, gray, greenish
Luster: Varies
Streak: Varies
Transparency: Transparent, opaque

Description: Fulgurites are formed when lightning strikes easily melted rock, forming a crust or hollow tubes. Fulgurites are most often found in deserts or on beaches where lightning hits the sand. Its name is derived from the Latin *fulgur*, meaning "lightning."

North American Locations: Mainly in deserts, beaches, and sand dunes across North America, but also wherever lightning can strike and melt sand.

Collecting Tips: Fulgurites are basically little tubes of sand and natural glass, and therefore are extremely fragile. Padding (e.g., Bubble Wrap) and storage containers should be brought along on any fulgurite-finding adventure.

Processing Tips: Fulgurites are collected, dusted off, and put on display just as they are found. They are far too fragile for any sort of lapidary or craft.

GALENA

Category: Sulfide
Crystal System: Cubic
Crystal Habit: Cubic
Cleavage: Perfect in three directions
Fracture: Subconchoidal

Mohs Scale: 2.5
Color: Gray
Luster: Metallic
Streak: Gray
Transparency: Opaque

Description: Galena is a lead sulfide and one of the more important lead ores mined. In fact, there was a lesser-known California "lead rush" before the famous California gold rush. Its name is derived from the Greek *galene*, meaning "lead ore."

North American Locations: Canada: Mountains, hills, and forests of British Columbia. United States: Arizona, Colorado, Idaho, Missouri, Oklahoma, and Wisconsin.

Collecting Tips: Galena occurs in veins in igneous rock and is recognized in the field by its bright metallic luster and heavy weight. Care should be taken when collecting galena, as it contains lead. Wear gloves while collecting the material and wash your hands afterward. If you plan on doing a lot of digging, wear a mask or respirator.

Processing Tips: Galena is soft, brittle, and full of lead, so it is not commonly used as a lapidary material. Fine crystals and specimens should be carefully cleaned and put on display out of the reach of small children and pets.

GARNET

Category: Nesosilicate
Crystal System: Cubic
Crystal Habit: Dodecahedral, trapezohedral
Cleavage: None
Fracture: Uneven, conchoidal

Mohs Scale: 6.5–7.5
Colors: Any, most commonly red tones
Luster: Vitreous
Streak: White
Transparency: Transparent, translucent, opaque

Description: Garnets are a large family of minerals in a wide range of colors—mostly red to purple tones. An inclusion of rutile in the crystal can create a star effect when cut. The word *garnet* is Middle English for "dark red." Garnet is the birthstone for January.

North American Locations: Mountains, forests, and river gravels, and along lake and ocean shorelines across North America. Notable collecting sites are in California, Idaho (placer deposits), Montana, New Jersey, and New York, and Quebec, Canada.

Collecting Tips: Garnet is found in metamorphic or igneous rock and is easily recognized by its crystal shape. It is heavy enough that it can be panned like gold. Look for weathered material that has worked its way to water from deposits in mountains, forests, and deserts.

Processing Tips: Specimens in matrix make for beautiful displays. The host rock can be carefully chiseled away to expose the crystal shape. Garnet is a popular stone for faceting and can also be turned into cabochons and tumbled.

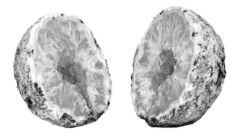

GEODE

Rock Type: Sedimentary, igneous
Structure: Hollow, generally spherical bodies often containing crystals on interior surfaces
Major Minerals: Quartz, calcite, celestine

Minor Minerals: Hematite
Fossils: None
Colors: Widely variable
Texture: Rough, earthy exterior; smooth, vitreous interior

Description: Geodes are generally round rocks with a hollow cavity containing one or more varieties of crystallized minerals. Most geodes are filled with colorless quartz crystals, but they can also contain amethyst, citrine, agate, and many others. They get their name from the Greek *geōdēs*, meaning "earthlike."

North American Locations: Mountains, forests, and plains, and along lake and ocean shorelines across North America. Notable collecting localities are in Oregon, Colorado, Arizona, and New Mexico.

Collecting Tips: Being hollow cavities, geodes are often much lighter than other rocks of the same general size. That said, some geodes can be completely filled with crystals and a bit trickier to identify without breaking or sawing them open.

Processing Tips: Geodes can be simply cracked open with a hammer. Chisels can be used to help crack the geode into two pieces. Geodes can also be cut with a rock saw, and the cut faces can be polished. Flatter pieces with small crystals can be turned into cabochons.

GOETHITE

Category: Hydroxide
Crystal System: Orthorhombic
Crystal Habit: Prismatic, elongated
Cleavage: Perfect in one direction
Fracture: Uneven to splintery
Mohs Scale: 5–5.5

Colors: Orange, brown, black
Luster: Adamantine, metallic
Streak: Brownish yellow, red
Transparency: Translucent, opaque

Description: Goethite is a type of hydrous iron oxide. It can often be found as a pseudomorph of other iron ore minerals such as pyrite and siderite. It was named after Johann Wolfgang von Goethe, a German poet, author, and mineralogist.

North American Locations: All geological settings across North America.

Collecting Tips: Goethite can be found in iron ore deposits and sometimes makes a weathered cap called an "iron hat." Care should be taken when extracting goethite from its host rock so as not to damage the fragile crystals. As an iron ore, it can be messy to handle, so wear clothes you don't mind getting rust-stained.

Processing Tips: Goethite is rarely used as lapidary material but can look great as an inclusion in other minerals used for lapidary uses, such as quartz. Specimens and crystals are cleaned, prepped, and put on display.

GOLD

Category: Native element
Crystal System: Cubic
Crystal Habit: Octahedral, dodecahedral, dendritic
Cleavage: None
Fracture: Hackly

Mohs Scale: 2.5–3
Color: Golden yellow
Luster: Metallic
Streak: Golden yellow
Transparency: Opaque

Description: Gold occurs as a soft, heavy metal golden yellow to pale yellow in color. The name has evolved over the years from the Gothic *gulth* and later the Old English *geolu*, both meaning "yellow."

North American Locations: Canada: Mountains and river gravels in British Columbia and Yukon Territory. United States: Mountains and river gravels in Georgia, North and South Carolina, and all western states and the Alaska shoreline.

Collecting Tips: Gold is found in individual particles and nuggets in placer (river) deposits, and in matrix rock in lode deposits. It is often found with black magnetite sands. Pyrite is commonly mistaken in the field for gold. They have similar yellow metallic coloring, but if struck with a geology pick, pyrite will shatter, while gold will flatten.

Processing Tips: Many gold miners keep nuggets and fine gold flakes in small vials and put them on display or into safe storage. Gold can be melted down for ingots or to be used to make jewelry and other decorative crafts. Wire and crystalline gold are considered more valuable as specimens than melted and processed gold and are generally put into displays.

GOSHENITE

Category: Cyclosilicate
Crystal System: Hexagonal
Crystal Habit: Prismatic
Cleavage: Indistinct in one direction
Fracture: Uneven to conchoidal

Mohs Scale: 7.5–8
Color: Colorless
Luster: Vitreous
Streak: White
Transparency: Transparent, translucent

Description: Goshenite is the colorless variety of the beryl family. It is named for Goshen, Massachusetts, where it was first discovered in the United States.

North American Locations: Canada: Mountains in Ontario. United States: Mountains in Colorado, Massachusetts, and South Dakota.

Collecting Tips: Goshenite is found in pegmatites and is best recognized in the field by its colorless hexagonal crystals. Care must be taken when extracting it so as not to damage the rare gemstone.

Processing Tips: Goshenite, like all beryls, makes for beautiful faceted gemstones to be set in jewelry. It can also be cut en cabochon, carved, tumbled, or simply put on display as is. Crystals with any stubborn staining can be cleaned with mild acids.

GRANITE

Rock Type: Felsic, plutonic, igneous
Structure: Crystalline
Major Minerals: Feldspar, quartz, mica

Minor Minerals: Biotite, hornblende
Fossils: None
Colors: White, gray, pink, red, black
Texture: Medium- to coarse-grained

Description: Granite is the most common intrusive igneous rock in the continental crust. Granite is host to pegmatites, which sometimes contain gemstones and rare minerals. Granite is commonly mined in massive quantities as a reliable building material.

North American Locations: Bedrock of many mountainous regions across North America.

Collecting Tips: Loose, fragmented specimens of granite can be found wherever granite bedrock is exposed.

Processing Tips: Granite specimens can be cleaned with water and a stiff brush.

GYPSUM

Category: Sulfate
Crystal System: Monoclinic
Crystal Habit: Prismatic, tabular
Cleavage: Perfect in one direction, distinct in two others
Fracture: Splintery, conchoidal

Mohs Scale: 2
Colors: Colorless, white, brown, yellow, pink
Luster: Subvitreous, pearly
Streak: White
Transparency: Transparent, translucent

Description: Gypsum is a type of calcium sulfate hydrate that comes in many forms. Transparent crystals of gypsum are called selenite. Massive fine-grained gypsum is called alabaster. It has been used for mortar and plaster since the time of ancient Egypt and derives its name from the Greek *gypsos*, meaning "plaster."

North American Locations: Mountains, deserts, forests, dry lake beds, and along some ocean shorelines across North America. Notable occurrences are in New Mexico (the "sand" dunes in White Sands National Park are pure gypsum), New York, Ohio, Oklahoma, and Utah, and in New Brunswick, Canada.

Collecting Tips: Gypsum is generally found in beds formed by evaporated ocean brine. It is soft and can be easily damaged by common rock tools. Take care when extracting so as not to damage crystals. Alabaster can be broken or cut out in chunks suitable for carving.

Processing Tips: Selenite crystals can be damaged by fresh water and even completely dissolve in it over time. A dry, soft toothbrush can be used to clean out dust and loose dirt. For more stubborn stains, some rockhounds use isopropyl alcohol.

HALITE

Category: Halide
Crystal System: Cubic
Crystal Habit: Cubic
Cleavage: Perfect in three directions
Fracture: Conchoidal

Mohs Scale: 2.5
Colors: Colorless, white, pink, orange, blue, purple
Luster: Vitreous
Streak: White
Transparency: Transparent, translucent

Description: Halite is more commonly known as salt. Yes, this sodium chloride is the same thing you keep in your kitchen. It derives its name from the Greek word for "salt," *halos.*

North American Locations: Mountains and forests, and especially in the dry lake beds of deserts and plains across North America. Excellent collecting areas are located in Oklahoma, New Mexico and California. The world's largest salt mine, located in Goderich, Ontario, extends beneath Lake Huron.

Collecting Tips: Halite is found in evaporated saline deposits. Crystals are very fragile and soluble in fresh water. Collecting salt minerals can dry out your hands very quickly, so wear gloves.

Processing Tips: Much care must be taken when displaying and storing halite. It is water-soluble. High and low humidity especially affects the mineral and can make it disintegrate.

HEMATITE

Category: Oxide
Crystal System: Hexagonal
Crystal Habit: Tabular, platy, botryoidal
Cleavage: None
Fracture: Uneven to splintery

Mohs Scale: 5–6
Color: Steelgray
Luster: Metallic, dull
Streak: Cherry red, reddish brown
Transparency: Opaque

Description: Hematite is a common iron oxide with a high iron content. Its name derives from the Greek *haimatites*, which means "bloodlike," a reference to the mineral's red streak. Purified ground hematite, called rouge, is used as a polish for glass, jewelry, and gems.

North American Locations: All geological settings across North America. Large hematite deposits are located in Michigan, Minnesota, New York, Utah, and Alabama, and in Ontario and Newfoundland and Labrador, Canada. The sandstone in the "red-rock" country of Utah is colored by hematite.

Collecting Tips: Hematite is common in sedimentary beds and metamorphosed sediments. It can often be found oxidized, with a red hue. Hematite is very heavy due to its high iron content.

Processing Tips: Hematite takes a great polish and can be cut into cabochons, beads, and carvings. Hematite's red streak isn't limited to test plates; the mineral will also stain the cutting lubricant in rock saws, turn cabochon wheels the same color, and stain hands and clothes.

HEMIMORPHITE

Category: Sorosilicate
Crystal System: Orthorhombic
Crystal Habit: Asymmetric, botryoidal, prismatic, tabular, massive, stalactitic

Cleavage: Perfect in one direction
Fracture: Uneven to conchoidal
Mohs Scale: 4.5–5
Colors: Colorless, white, yellow, blue, green, pink

Luster: Vitreous, adamantine, pearly
Streak: White
Transparency: Transparent, translucent

Description: Hemimorphite is a hydrated zinc silicate and derives its name from the Greek *hemi*, meaning "half," and *morph*, meaning "form." This is a reference to the mineral's crystal habit.

North American Locations:
Canada: Mountains and hills of British Columbia and Ontario. United States: Mountains and hills of Idaho, Colorado, Missouri, New Jersey, and Tennessee.

Collecting Tips: Hemimorphite is formed as a secondary mineral in alteration zones of zinc deposits. Botryoidal hemimorphite is recognized in the field by its crystal habit and weight. It is a bit heavier than the average rock. It can also be recognized by its bright pastel colors. Crystals can be mistaken for calcite. Care should be taken when extracting crystals.

Processing Tips: Massive hemimorphite can be carved, tumbled, and turned into cabochons. Rare transparent, colorless specimens can be faceted. Most hemimorphite is collected as specimens. The matrix can be prepped before displaying.

HERDERITE

Category: Phosphate
Crystal System: Monoclinic
Crystal Habit: Prismatic, tabular
Cleavage: Fair in one direction
Fracture: Subconchoidal
Mohs Scale: 5–5.5

Colors: Colorless, yellow, greenish
Luster: Vitreous
Streak: White
Transparency: Transparent, translucent

Description: Herderite, a calcium beryllium phosphate, was named after German mining official August Wolfgang von Herder. It is collected for its well-formed, stubby, pyramidal crystals in pale shades of yellow and green.

North American Locations: Canada: Mountains in Manitoba. United States: California, Maine, New Hampshire, North Carolina, and South Dakota.

Collecting Tips: Herderite is an uncommon mineral found in granite pegmatites. It can be recognized in the field by its generally prismatic crystal structure. Care should be taken extracting these fragile crystals from their pegmatite matrix, as they can be easily damaged. Herderite is often found with albite feldspar, quartz, topaz, and tourmaline.

Processing Tips: Rare transparent crystals can be faceted into beautiful collectors' gems. Most herderite is collected as mineral specimens, but damaged crystals may be turned into cabochons. Matrix specimens can be trimmed and prepped for better display.

HEULANDITE

Category: Zeolite
Crystal System: Monoclinic
Crystal Habit: Tabular
Cleavage: Perfect in one direction
Fracture: Uneven
Mohs Scale: 3.5–4

Colors: Colorless, white, gray, yellow, pink, red, brown
Luster: Vitreous, pearly
Streak: Colorless
Transparency: Transparent, translucent

Description: Heulandite is a member of the large family of zeolite minerals and is very similar to stilbite. In fact, it was confused with stilbite until 1822 when chemist H. J. Brooke distinguished it as a separate species and named it *heulandite* after mineralogist John Henry Heuland.

North American Locations: Canada: Mountains and hills of Nova Scotia. United States: Colorado, New Jersey, and Oregon.

Collecting Tips: Heulandite is found in many deposits, including those of granite and its pegmatites, basalt, andesite, diabase, and metamorphic rocks. It can be recognized in the field by its crystal habit. Care must be taken when extracting crystals and clusters so as not to damage them.

Processing Tips: Most specimens collected in the field are used for display pieces and can be trimmed or prepped for a more aesthetic presentation. Dirty specimens can usually be cleaned with a soft toothbrush, soap, and water.

HORNBLENDE

Category: Inosilicate
Crystal System: Monoclinic
Crystal Habit: Prismatic
Cleavage: Fair in two directions
Fracture: Uneven
Mohs Scale: 5–6

Colors: Gray, black
Luster: Vitreous
Streak: White, gray
Transparency: Translucent, opaque

Description: Hornblende is the name used for a group of minerals only distinguishable by chemical analysis. The name is derived from the German *horn*, meaning "dark ore of no value," and *blenden*, meaning "to blind or confuse." Hornblende can often look like valuable metal ore.

North American Locations: All geological settings across North America. Fine specimens come from Arizona, California, Delaware, New Jersey, and Pennsylvania; the best Canadian specimens come from British Columbia, Nova Scotia, and Ontario.

Collecting Tips: Hornblende can be found in metamorphic rock such as amphibolite, schist, gneiss, and also mafic igneous rock. Crystals can range from microscopic to a few inches long and are generally recognized by their black color and prismatic structure. Care should be taken when extracting rare, fine crystals.

Processing Tips: Rock with hornblende crystals can be cut, carved, and turned into cabochons. Rare, fine crystal specimens are generally displayed in matrix and sometimes trimmed for better display. Fine crystals are too brittle to cut into cabochons.

HYPERSTHENE

Category: Inosilicate
Crystal System: Orthorhombic
Crystal Habit: Coarsely crystalline, massive
Cleavage: Perfect in one direction
Fracture: Uneven

Mohs Scale: 5.5–6
Colors: Gray, brown, green, purple
Luster: Vitreous, pearly
Streak: Gray, greenish white
Transparency: Transparent, opaque

Description: Hypersthene, an orthorhombic pyroxene, is a common rock-forming mineral. It derives its name from the Greek *hyper*, meaning "above," and *stenos*, meaning "power." This is in reference to it being harder than hornblende, a closely related mineral it is often confused for.

North American Locations: All geological settings across North America. Fine specimens come from the mountains of California, Colorado, North Carolina, Oregon, and Pennsylvania, and of Quebec, Canada.

Collecting Tips: Hypersthene is found in igneous and metamorphic rocks. Fine specimens often exhibit a mauve labradorescence.

Processing Tips: Hypersthene is a popular mineral for cutting en cabochon and using in jewelry. The mineral can also be tumbled and carved for various lapidary purposes.

INOCERAMUS

Category: Fossil
Mohs Scale: 3–4
Colors: Gray, tan, brown; mother-of-pearl coating has wide range of iridescent colors

Luster: Earthy, dull; mother-of-pearl coating is vitreous to waxy
Transparency: Opaque; mother-of-pearl coating is translucent

Description: *Inoceramus* is an extinct genus of bivalve mollusks called pelecypods (clams) that thrived in ancient seas. The name comes from the Greek word for "strong pot," alluding to the strength and thickness of the shell. *Inoceramus* fossils are often adorned with gleaming coatings of mother-of-pearl.

North American Locations: Plains, forests, and along riverbanks in central North America. Notable collecting areas are in Colorado, Kansas, Montana, Nebraska, and South Dakota, and in Alberta and Saskatchewan, Canada.

Collecting Tips: Search for exposures of soft gray shale along the banks of creeks and rivers. These fossils can often be collected from the surface, but some digging may be required.

Processing Tips: *Inoceramus* fossils should be cleaned with water and a soft brush. Exercise caution when cleaning, as the mother-of-pearl coatings, if present, can be fragile.

JADE–JADEITE

Category: Pyroxene
Crystal System: Monoclinic
Crystal Habit: Massive, fibrous, prismatic
Cleavage: Good in two directions
Fracture: Uneven to splintery

Mohs Scale: 6.5–7
Color: Any
Luster: Greasy, vitreous
Streak: Colorless
Transparency: Transparent, translucent

Description: Jade has long been coveted by humans, and archaeological evidence has shown that it has been used for tools and weapons for at least one hundred thousand years. Jadeite is a pyroxene mineral and one of two minerals recognized as jade.

North American Locations: Canada: In river and shoreline gravels in Alberta and Newfoundland and Labrador. United States: Along the central California ocean shoreline.

Collecting Tips: Jadeite has a granular composition of tough, interlocking crystals as opposed to nephrite's more fibrous structure. It is most often found in green tones, but other colors can also be found in the field.

Processing Tips: Jadeite takes a wonderful polish. This resilient material is very strong, but is easily carved, shaped, and tumbled, and makes for superb cabochons that are excellent for setting into all types of jewelry.

JADE—NEPHRITE

Category: Inosilicate
Crystal System: Monoclinic
Crystal Habit: Massive
Cleavage: None
Fracture: Splintery
Mohs Scale: 5–6

Color: Any
Luster: Dull, waxy
Streak: White
Transparency: Translucent, nearly opaque

Description: Nephrite is an inosilicate and one of two minerals recognized as jade. It has a more fibrous structure than jadeite, which is grainier. It derives its name from the Greek *nephriticus*, meaning "kidney stone," a reference to its once being believed to be a cure for kidney stones and kidney ailments.

North American Locations: Canada: Along the ocean shorelines of British Columbia. United States: Along the ocean shorelines of Alaska, California, Oregon, and Washington. It is also found in the hills and stream gravels of Wyoming.

Collecting Tips: Nephrite is generally green in color but can be many other colors. Jade tends to occur with other greenish metamorphic minerals; it can also form a tough crust, making identification more difficult. Light will shine through nephrite and a steel blade won't scratch it.

Processing Tips: Nephrite takes a wonderful polish. This material is very strong, but is easily carved, shaped, and tumbled, and makes for superb cabochons that are excellent for setting into all types of jewelry.

JASPER

Category: Tectosilicate
Crystal System: Hexagonal
Crystal Habit: Cryptocrystalline
Cleavage: None
Fracture: Conchoidal
Mohs Scale: 7

Color: Any
Luster: Vitreous
Streak: White
Transparency: Opaque

Description: Jasper is an opaque variety of microcrystalline quartz. It has long been used to make tools, weapons, and ornamental items. The name is derived from the Greek *iaspis*, which was a term used in ancient Greece for almost any brightly colored stone. It can often have bands of quartz or agate running through it.

North American Locations: All geological settings across North America.

Collecting Tips: Jasper is typically found in metamorphic, igneous, and sedimentary deposits. It is recognized in the field by its vitreous or waxy luster, conchoidal fracture, and opaque transparency. It can come in any color or combination of colors, but is most commonly found in red, green, and brown tones.

Processing Tips: Jasper is hard and takes an excellent polish. It is perfect for tumbling, carving, and making cabochons or a polished face cut to display the variety of patterns within the mineral.

Rock	**JET**

Rock Type: Sedimentary, altered
Structure: Solid, amorphous
Major Minerals: Carbon
Minor Minerals: Sulfur, quartz, hematite
Fossils: None

Colors: Jet black with occasional brown streaks
Texture: Very fine-grained

Description: Jet, a form of lignite coal, is jet black in color, very light in weight. Its name derives from *jaiet*, the French word for the same material. Jet jewelry was popular in ancient Rome and later during the Victorian era.

North American Locations: Canada: No known deposits. United States: On plains and along rivers and shorelines in Colorado, Maryland, New Mexico, and Utah.

Collecting Tips: Jet is most often found in gravels and sediments, where it can be identified by its deep black color. Because of its lightness (it is almost light enough to float in water), it is the first material to separate when gravels are washed.

Processing Tips: Clean jet only with soap, water, and a soft brush, then polish with a jeweler's polishing cloth. Avoid contact with any hard materials. Jet takes a fine polish and is easily carved into beads and other ornamental objects.

KORNERUPINE

Category: Borosilicate
Crystal System: Orthorhombic
Crystal Habit: Prismatic, massive, fibrous
Cleavage: Fair in one direction
Fracture: Uneven

Mohs Scale: 6–7
Colors: Colorless, white, gray, green, blue, brown, black
Luster: Vitreous
Streak: White
Transparency: Transparent, translucent

Description: Also known as prismatine, kornerupine is a rare type of borosilicate. It was first discovered in Greenland in 1884 by Andreas Nikolaus Kornerup, for whom it is named.

North American Locations: Canada: Forests and mountains in Ontario and Quebec. United States: Forests and mountains in New Jersey, New York, and Utah.

Collecting Tips: Kornerupine is found in boron-rich metamorphic and volcanic rock. It can be recognized in the field by its prismatic crystal habit, variety of color, and hardness. It can be easily confused with many other minerals, and further chemical and physical testing may be needed to confirm its identification.

Processing Tips: Kornerupine is most commonly collected as a specimen. Crystals in matrix may be trimmed or prepped for better display. Very rare, fine translucent crystals can be faceted into gems and are hard enough for most jewelry applications.

KYANITE

Category: Nesosilicate
Crystal System: Triclinic
Crystal Habit: Columnar, fibrous, bladed
Cleavage: Perfect in one direction
Fracture: Splintery

Mohs Scale: 4–5 parallel to one axis; 6.5–7 perpendicular to same axis
Colors: Blue, white, green, gray, yellow, pink, orange, black
Luster: Vitreous, pearly

Streak: Colorless
Transparency: Transparent, translucent

Description: Kyanite is an aluminum-silicate mineral that derives its name from the Greek *kyanos*, meaning "blue," a reference to the mineral's common blue tone. Kyanite is a raw material that can be used industrially in ceramics, electronics, spark plugs, and abrasives.

North American Locations: Canada: Mountains and forests of New Brunswick. United States: Mountains and forests of Georgia and New York.

Collecting Tips: Kyanite is found in outcrops of metamorphosed schists and gneisses as well as granitic pegmatites. It is most commonly found in cyan-blue tones, especially in North America. Care must be taken when extracting specimens so as not to damage them.

Processing Tips: Only experienced lapidaries should attempt cutting and polishing this finicky mineral. That said, translucent specimens can make wonderful cabochons and stunning faceted gems. Most specimens are kept as found.

LABRADORITE

Category: Tectosilicate
Crystal System: Triclinic
Crystal Habit: Massive, rare tabular
Cleavage: Perfect in two directions
Fracture: Uneven to conchoidal
Mohs Scale: 6–6.5

Colors: Iridescent blue, green, gold, and magenta
Luster: Vitreous
Streak: White
Transparency: Translucent, transparent

Description: Labradorite derives its name from Canada's Labrador Peninsula, where it was first discovered. The mineral subsequently became the namesake for the optical effect it displays of iridescent colors known as labradorescence. A unique variety of yellow labradorite colored red, blue, and green can only be found in Oregon.

North American Locations: Canada: Along the shoreline of Newfoundland and Labrador. United States: Mountains in Arizona, California, New Mexico, North Carolina, and Oregon.

Collecting Tips: Labradorite can be found in igneous, metamorphic, and sedimentary rocks. Gem-quality labradorite can be recognized by its labradorescence. Yellow labradorite is generally a transparent champagne yellow and looks like broken glass.

Processing Tips: Labradorite is an excellent lapidary material and takes a great polish. It can be used for carving, tumbling, faceting, and making cabochons. It is hard enough to set into all types of jewelry. Care should be taken when grinding specimens, as the labradorescence can be easily ground away.

Rock	**LAPIS LAZULI**

Rock Type: Metamorphic
Structure: Microcrystalline, massive
Major Minerals: Lazurite
Minor Minerals: Pyrite, calcite
Fossils: None

Colors: Light blue to midnight blue
Texture: Very fine-grained

Description: Lapis lazuli has been dug for over five thousand years. It is commercially mined in Afghanistan and Chile, but a few deposits exist in North America. Lapis is a rock consisting of the minerals lazurite, calcite, and pyrite. It derives its name from both the Latin *lapis*, meaning "stone," and the Persian *lazhward*, meaning "blue."

North American Locations: Canada: Mountains in Nunavut. United States: Mountains in Colorado and California.

Collecting Tips: Lapis is easily recognized in the field by its deep to bright blue tones, mottled with white and brassy specks. Lapis forms as a large mass, so hard-rock mining tools are necessary to remove specimens suitable for lapidary purposes.

Processing Tips: Lapis takes a wonderful polish and can be tumbled, carved, and turned into cabochons that are very popular set in all types of jewelry.

LEPIDOLITE

Category: Phyllosilicate
Crystal System: Monoclinic
Crystal Habit: Tabular, prismatic, pseudohexagonal, scaly aggregates, massive
Cleavage: Perfect in one direction

Fracture: Uneven
Mohs Scale: 2.5–3
Colors: Pink, purple, colorless, gray
Luster: Vitreous, pearly
Streak: Colorless
Transparency: Transparent, translucent

Description: Lepidolite, a variety of mica, is the earth's most common lithium-bearing mineral. It derives its name from the Greek *lepidos*, meaning "scale," and *lithos*, meaning "stone"—a reference to the mineral's flaky crystal structure.

North American Locations: Canada: Mountains in Ontario and Quebec. United States: Mountains in Colorado, California, New Mexico, Maine, and South Dakota.

Collecting Tips: Lepidolite is found almost solely in granitic pegmatite deposits. It is recognized in the field by its pink and purple tones. It's a mica, so expect to get all sorts of "nature's glitter" on your skin, clothes, and vehicle. Lepidolite is commonly found with tourmaline and quartz.

Processing Tips: Most lepidolite is not suitable for lapidary purposes and instead is collected as a mineral specimen as found. Quartz with lepidolite inclusions make excellent lapidary material for carving, tumbling, and making cabochons that can be used in jewelry.

LIMESTONE

Rock Type: Marine, chemical, sedimentary
Structure: Usually bedded
Major Minerals: Calcite
Minor Minerals: Aragonite, dolomite, siderite, quartz, pyrite

Fossils: Marine and freshwater invertebrates
Colors: Mostly gray; also off-white, brown, tan, and pinkish
Texture: Fine- to medium-grained, angular, rounded

Description: Limestone is a rock mostly made out of calcite. It usually forms in warm, shallow seas from either the precipitation of calcium carbonate from water or the accumulations of the calcareous shells and skeletons of marine life. Limestone is mined commercially for use in building materials, cement, glassmaking, and agriculture.

North American Locations: All geological settings across North America.

Collecting Tips: Limestone is recognizable in the field by its fine-grained texture. Often fossils can be found in limestone deposits. Hard-rock mining equipment is necessary if you plan on removing specimens from an outcrop.

Processing Tips: Limestone is of little lapidary use unless fossil specimens can be seen with the naked eye. Specimens with interesting fossils can be carved and cut en cabochon, although it's not recommended for most jewelry.

LIMONITE

Category: Hydroxide
Crystal System: Mixture
Crystal Habit: Massive, oolitic, stalactitic
Cleavage: Splintery to conchoidal
Fracture: Uneven

Mohs Scale: 5–5.5
Colors: Brown, yellow
Luster: Earthy, submetallic, dull
Streak: Yellow brown
Transparency: Opaque

Description: *Limonite* is a term used for unidentified iron oxides and hydroxides. Most limonite is considered to be goethite, and often the names are used interchangeably. The name is derived from the Greek *leimōn*, meaning "wet meadow." This is a reference to the mineral occurring as a bog iron ore in marshes and meadows.

North American Locations: All geological settings across North America.

Collecting Tips: Limonite is easily recognized in the field by its bright, warm color tones. It looks a lot like rust.

Processing Tips: Limonite is generally collected by rockhounds as a mineral specimen and placed on display. In its raw form, limonite is of little lapidary use unless it is an inclusion within colorless quartz, which can be carved, tumbled, cut en cabochon, and even faceted.

MAGNETITE

Category: Oxide
Crystal System: Cubic
Crystal Habit: Octahedral
Cleavage: None
Fracture: Uneven to subconchoidal
Mohs Scale: 5.5–6

Colors: Black, brown
Luster: Metallic, semimetallic
Streak: Black
Transparency: Opaque

Description: Magnetite is a variety of iron oxide that falls in the spinel family of minerals. It is often assumed that it derives its name from its magnetic properties, but it was actually named after the Magnesia region in Greece. Even today it is the most commonly mined iron ore.

North American Locations: All geological settings across North America. Fine specimens come from Alabama, California, Minnesota, New York, and Utah, and Ontario, Canada.

Collecting Tips: Magnetite can be found in igneous, metamorphic, and sedimentary rocks. It is easily identifiable in the field because it is highly magnetic. Crystals are identified by their octahedral shape.

Processing Tips: While it is common and hard and takes a good polish, magnetite is not commonly used for lapidary purposes. Some specimens, however, can be tumbled, carved, and cut en cabochon. Fine crystals can be cleaned with a light soap and soft toothbrush.

MALACHITE

Category: Carbonate
Crystal System: Monoclinic
Crystal Habit: Massive, botryoidal
Cleavage: Perfect in one direction
Fracture: Splintery to subconchoidal

Mohs Scale: 3.5–4
Color: Green
Luster: Adamantine, silky
Streak: Pale green
Transparency: Opaque, translucent

Description: Malachite is a copper carbonate hydroxide and is thought to be the earliest known mined copper ore. In ancient times it was used as a paint pigment, eye makeup, and glaze for glass. It derives its name from the Greek *malache*, meaning "mallow"—a reference to the mineral's green tone.

North American Locations: Although small amounts of malachite can be found in mountains across North America, it is most common in the deserts of Arizona and New Mexico.

Collecting Tips: Malachite can be found in the altered zones of copper deposits. The bright green tones of malachite are one of the first signs rockhounds will notice. Malachite is soft and can be easily scratch-tested.

Processing Tips: While fairly soft, malachite is a popular lapidary material. It can be easily carved and makes beautiful ornamental items. Cabochons are popular in jewelry. Caution should be taken when carving or grinding malachite, as the copper dust is harmful to breathe.

| Rock | **MARBLE** |

Rock Type: Regional, contact metamorphic
Structure: Microcrystalline, massive
Major Minerals: Calcite

Minor Minerals: Diopside, tremolite, actinolite
Fossils: None
Colors: White, pink, black, gray, green, blue
Texture: Fine-grained

Description: Marble is a metamorphic rock that consists of altered limestone or dolomite. It derives its name from the Greek *mármaron*, meaning "crystalline rock," and *mármaros*, meaning "shining stone." Marble has been used since ancient times for carving statues and monuments, and for decorating buildings.

North American Locations: Mountains across North America. Notable deposits are in California, Colorado, New Hampshire, New York, and Vermont, and in Ontario and Quebec, Canada.

Collecting Tips: Marble can be collected as talus material. Look for pieces that are suitable for lapidary projects. Hard-rock mining equipment will be needed to extract large pieces from the marble deposit. Make sure the material you're collecting doesn't have any noticeable cracks.

Processing Tips: It can be carved, tumbled, and even cut en cabochon, but the rock is considered too soft for most jewelry applications. Marble can be carved for both ornamental and utilitarian purposes.

MESOLITE

Category: Tectosilicate
Crystal System: Monoclinic
Crystal Habit: Acicular
Cleavage: Perfect in two directions
Fracture: Uneven
Mohs Scale: 5

Colors: Colorless, white
Luster: Vitreous, silky
Streak: White
Transparency: Transparent, translucent

Description: Mesolite is a part of the zeolite family and is closely related to natrolite, which it can also resemble and be mistaken for. In fact, it derives its name from the Greek *meso*, meaning "middle," a reference to the mineral's composition falling between natrolite and scolecite.

North American Locations: Forests, mountains, and deserts across North America. Fine specimens come from California, Colorado, and New Jersey, and from Nova Scotia, Canada.

Collecting Tips: Mesolite occurs in vugs found in basalt, andesine, and also in hydrothermal veins. It is recognized in the field by its fibrous sprays of crystals. It may be mistaken for natrolite, and sometimes only chemical analysis can tell the two apart. Care must be taken when extracting extremely fragile crystals from host rock.

Processing Tips: Mesolite crystals are very fragile and the utmost care should be taken when prepping specimens for display. Crystals may be cleaned with distilled water and a soft brush. It is possible to facet large crystals for collectors.

MIMETITE

Category: Arsenate
Crystal System: Hexagonal
Crystal Habit: Prismatic, acicular, reniform, botryoidal, globular
Cleavage: Fair in one direction
Fracture: Uneven to subconchoidal
Mohs Scale: 3.5–4

Colors: Pale yellow, yellow brown, green, orange
Luster: Resinous
Streak: White
Transparency: Transparent to translucent

Description: Mimetite is a lead arsenate chloride mineral. It derives its name from the Greek *mimethes*, meaning "imitator," a reference to the mineral being commonly mistaken for pyromorphite. Mimetite is also known as arsenopyromorphite, mimetesite, and prixite. There is also a barrel-shaped orange-red-brown variety called campylite.

North American Locations: Mountains and arid desert regions across western North America.

Collecting Tips: Mimetite is generally found as a secondary mineral in oxidized zones of lead deposits. Although it can easily be confused for pyromorphite, mimetite is recognized in the field by its yellow tones. Care should be taken extracting fragile crystals from host rock.

Processing Tips: Although mimetite can be faceted, it is generally considered too soft for most gemstone and lapidary purposes and is collected mostly as a mineral specimen. Care should be taken when prepping matrix specimens with crystals so as not to dislodge or break them.

MOLYBDENITE

Category: Sulfide
Crystal System: Hexagonal/ trigonal
Crystal Habit: Tabular, prismatic
Cleavage: Perfect in one direction
Fracture: Sectile

Mohs Scale: 1–1.5
Color: Lead gray
Luster: Metallic
Streak: Greenish, bluish gray
Transparency: Opaque

Description: Molybdenite is the most important ore of molybdenum, which is crucial in the production of specialty metal alloys. The rare metal is also used in fertilizers and batteries. It derives its name from the Greek *molybdos*, meaning "lead," a reference to its originally being identified as lead.

North American Locations: Mountains and deserts across western North America, notably in Colorado, New Mexico, and Idaho.

Collecting Tips: Molybdenite can be found in granite, pegmatites, hydrothermal veins, and contact metamorphic rocks. Crystals are especially soft and fragile, so extreme care must be taken when extracting the mineral from host rock and especially during transport.

Processing Tips: Molybdenite is collected as a specimen only, as it is much too soft for any lapidary purposes. Much care should be taken with display specimens, as the crystals can be easily damaged by even the lightest touch.

MORGANITE

Category: Cyclosilicate
Crystal System: Hexagonal
Crystal Habit: Tabular
Cleavage: Fair in one direction
Fracture: Uneven to conchoidal
Mohs Scale: 7.5–8

Colors: Pink, peach, orange, pinkish yellow
Luster: Vitreous
Streak: White
Transparency: Transparent, translucent

Description: Morganite is the pink- to peach-toned variety of the beryl family. It owes its unique colors to traces of cesium or manganese. Morganite was named after the American financier J.P. Morgan.

North American Locations: Canada: No known deposits. United States: Mountains in California and Maine.

Collecting Tips: Morganite is very rare but can be found in lithium-rich pegmatites and is often found with lepidolite and tourmaline. It is recognized in the field by its bright pink to peach tones and hexagonal crystal system. Hard-rock mining tools and a delicate touch are required when extracting fragile crystals from vugs.

Processing Tips: Morganite makes for beautiful, faceted gemstones to be set in jewelry. It can also be cut en cabochon, carved, tumbled, or put on display as is. Crystals with any stubborn staining can be cleaned with a light acid solution. Low-grade morganite can be heat-treated to improve its color.

MUSCOVITE

Category: Phyllosilicate
Crystal System: Monoclinic
Crystal Habit: Tabular, pseudohexagonal
Cleavage: Perfect in one direction
Fracture: Micaceous

Mohs Scale: 2.5–3
Colors: Colorless, white, pale green, pink, brown
Luster: Pearly, submetallic
Streak: Brownish white
Transparency: Transparent, translucent

Description: Muscovite is a common mica mineral. Large "books" of muscovite, sometimes up to several feet wide, can be split into transparent sheets. The mineral was mined in Russia in the 1700s for windowpanes and was called Muscovy glass.

North American Locations: Mountains across North America. Fine specimens come from California, Maine, Massachusetts, New Hampshire, North Carolina, and South Dakota, and from British Columbia and Quebec, Canada.

Collecting Tips: Muscovite is found as a common rock-forming mineral in igneous rocks, especially granites, and also in hydrothermal veins and pegmatites. Mica crystals are very soft and fragile, so extreme care must be taken when extracting them.

Processing Tips: Mica is not used for lapidary purposes, as it is too soft and cleaves too easily. Crystal specimens in matrix can be trimmed and prepped for display. Mica as an inclusion inside quartz can be used for lapidary purposes like tumbling and cutting cabochons.

NATROLITE

Category: Inosilicate
Crystal System: Orthorhombic
Crystal Habit: Acicular
Cleavage: Perfect in one direction
Fracture: Uneven
Mohs Scale: 5–5.5

Colors: Colorless, pink, white, gray, red, green, yellow
Luster: Vitreous, pearly
Streak: White
Transparency: Transparent, translucent

Description: A member of the large zeolite family, natrolite can often be confused with its close relatives mesolite and scolecite. Its name is derived from the Greek *nitron*, meaning "sodium," a reference to the mineral's sodium content.

North American Locations: Mountains in British Columbia, Nova Scotia, and Ontario. United States: Mountains in California, Colorado, New Jersey, and Oregon.

Collecting Tips: Natrolite occurs in vugs found in basalt and andesine, and also in hydrothermal veins. It is recognized in the field by its fibrous sprays of crystals. Care must be taken when extracting the extremely fragile crystals from host rock.

Processing Tips: Natrolite crystals are very fragile, and the utmost care should be taken when prepping specimens for display. Crystals may be cleaned with distilled water and a soft brush. Rare large crystals can be faceted for collectors.

NAUTILOID

Category: Fossil
Mohs Scale: 3–4
Colors: Gray, tan, brown

Luster: Earthy, dull
Transparency: Opaque

Description: Nautiloids are a large and diverse group of marine cephalopods. While nautiloids were once common throughout the world, the pearly nautilus is the only genus that still exists today. The first nautiloids had straight, conical shells but later developed whorled shells. Both types are found as fossils.

North American Locations: Canada: Hills, forests, plains, and riverbanks of the southern parts of central and eastern Canada. United States: Hills, forests, plains, and riverbanks of the central and northeastern United States.

Collecting Tips: Intact nautiloid fossils are easily removed from soft, weathered formations of limestone and shale. Removing these fossils from harder rock requires hammers and chisels. Great care must be taken not to damage the fossil.

Processing Tips: Nautiloid fossils are quite fragile and should be cleaned only with water and soft brushes. Bits of matrix rock can be removed by working slowly and carefully with knife points and dental picks.

NEPTUNITE

Category: Inosilicate
Crystal System: Monoclinic
Crystal Habit: Prismatic, tabular
Cleavage: Perfect in one direction
Fracture: Conchoidal
Mohs Scale: 5–6

Color: Black
Luster: Vitreous, submetallic
Streak: Cinnamon brown
Transparency: Opaque

Description: Neptunite is a rare and more recently discovered mineral, first being found in the early 1900s in Greenland. Its name is derived from the Roman sea god Neptune and is a nod to a similar-looking mineral called aegirine, which was named after a Norse sea god.

North American Locations: The main collecting site for neptunite is in the hills near Mont-Saint-Hilaire, Quebec, Canada. Specimens have also been found in the mountains of San Benito County, California.

Collecting Tips: Neptunite can be found in pegmatites, as well as in natrolite veins found in schist and serpentinite. Care must be taken when extracting exposed neptunite crystals from host rock.

Processing Tips: Neptunite is durable enough to be tumbled and cut en cabochon, but it's very rare and usually kept as a display specimen. It is often encased in natrolite and can be exposed or completely removed from the natrolite with an acid bath. It can also be faceted for collectors of faceted gems.

OBSIDIAN

Category: Mineraloid (volcanic glass)
Crystal Habit: Amorphous
Fracture: Conchoidal
Mohs Scale: 5–6

Colors: Black, gray, brown, red, orange, green
Luster: Vitreous
Streak: White
Transparency: Translucent, transparent, opaque

Description: Obsidian is a natural volcanic glass formed by the rapid cooling of highly siliceous magma. Its color is most commonly black, but the inclusion of other minerals can create other colors, as well as an optical effect known as sheen. It derives its name from Obsius, the Roman explorer who allegedly first found obsidian in Ethiopia.

North American Locations: Volcanic mountains across North America. Fine specimens have come from Arizona, California, Oregon, and Wyoming, and from British Columbia, Canada.

Collecting Tips: Wear hand and eye protection when collecting obsidian, as it is extremely sharp. Obsidian is easily identified by its glasslike appearance, usually black coloration, and sharp conchoidal fracture. Search for it as float, or remove chunks from a deposit.

Processing Tips: Obsidian can be used for tumbling, carving, knapping, cutting cabochons, and faceting. It is a popular material for flaking into arrowheads. Obsidian chips easily in a tumbler; it is recommended to use a buffing medium with your tumbler load.

Mineral | **OKENITE**

Category: Inosilicate
Crystal System: Triclinic
Crystal Habit: Fibrous, bladed
Cleavage: Perfect in one direction
Fracture: Splintery
Mohs Scale: 4.5–5

Colors: Colorless, white, yellow, blue
Luster: Pearly, vitreous
Streak: White
Transparency: Transparent, translucent

Description: Okenite is a calcium silicate hydrate and is quite often found with zeolite minerals, although it is not a member of this family. Its crystals can often be quite fuzzy, resembling the soft fur of a mammal. It was named after German naturalist Lorenz Oken in 1828.

North American Locations: Mountains across North America.

Collecting Tips: Okenite is often found in vugs found in basalt, andesine, and also in hydrothermal veins. It is easily recognized in the field by its soft, fibrous sprays of crystals. Take care, as the crystals are very fragile. Wrap specimens in tissue or Bubble Wrap for transportation.

Processing Tips: Okenite crystals are very fragile and the utmost care should be taken when prepping specimens for display. For prepping, avoid all chemicals and use nothing less delicate than a makeup brush.

OPAL

Category: Tectosilicate
Crystal Habit: Massive
Fracture: Conchoidal
Mohs Scale: 5–6
Colors: Vary widely; mostly iridescent

Luster: Vitreous, waxy
Streak: White
Transparency: Transparent, translucent

Description: Technically a mineraloid, opal is a hardened silica gel and typically contains 5–10 percent water. It derives its name from the Latin *opalus*, a Roman version of the Sanskrit word *upala*, which means "precious stone." Precious opal is rare and displays an iridescent optical phenomenon known as opalescence.

North American Locations: In river gravels in volcanic regions and in deserts and arid mountainous areas across North America. Notable collecting areas are in California, Nevada, and Oregon, and in British Columbia, Canada.

Collecting Tips: Precious opal is unmistakable in the field. Its opalescence will give it away every time. Common opal is recognized by its waxy to vitreous luster and conchoidal fracture. It's often softer to the touch than agate, jasper, or chalcedony.

Processing Tips: Common opal is easily tumbled, carved, cut en cabochon, and faceted. Precious opal should only be cut by experienced lapidaries. Opal can dry out easily and form fractures known as crazes. Some specimens are kept permanently in water to avoid this.

ORPIMENT

Category: Sulfide
Crystal System: Monoclinic
Crystal Habit: Massive, foliated
Cleavage: Good in one direction
Fracture: Sectile
Mohs Scale: 1.5–2

Colors: Yellow, orange
Luster: Resinous
Streak: Pale yellow
Transparency: Transparent, translucent

Description: Orpiment is an arsenic sulfide commercially mined for use in fireworks, semiconductors, photoconductors, pigments, linoleum, and the making of infrared-transmitting glass. It derives its name from the Latin *aurum*, meaning "gold," and *pigmentum*, meaning "paint," a reference to the mineral being used as a yellow pigment.

North American Locations: Mountains and deserts of western North America, notably in Nevada and Arizona.

Collecting Tips: Orpiment can be found in low-temperature hydrothermal veins, near hot springs, and as an alteration product of other arsenic-bearing minerals. It is recognized by its bright yellow to orange tones and foliated masses. Care should be taken when collecting orpiment due to its arsenic content. Wear a filtered face mask and gloves when collecting orpiment.

Processing Tips: Orpiment is generally collected as a mineral specimen. Care should be taken when cutting, grinding, or polishing orpiment due to its arsenic content. Wash your hands after handling. Keep it away from metals that can tarnish, as its sulfur content can speed up the oxidizing process.

PEARL

Category: Organic gem
Crystal Habit: Massive
Fracture: Uneven
Mohs Scale: 2.5–4
Colors: White, black, cream, yellow, blue, green, pink

Luster: Pearly
Streak: White
Transparency: Opaque

Description: A pearl is a type of concretion formed inside a mollusk's shell. Pearls can grow in both saltwater and freshwater mollusks, but the rarest are found in mollusks lined with nacre, or mother-of-pearl, on the inside of the shell. It derives its name from the Latin *perna*, meaning "ham, leg of pork" a reference to the shape of the mollusks most pearls are found in.

North American Locations: Bays, harbors, lakes, rivers, and streams, and along ocean shorelines across the United States and southern Canada. The most important collecting areas are the inland waterways of the southeastern United States.

Collecting Tips: You're going to have to open a lot of mollusks to find a pearl. It is generally suggested that you plan on cooking your found mollusks when searching so as not to waste them.

Processing Tips: Pearls are generally kept as found and not polished. They can be drilled to be used for jewelry and are especially popular as necklaces. Care should be taken when drilling so as not to breathe in the dust.

PECTOLITE

Category: Inosilicate
Crystal System: Triclinic
Crystal Habit: Acicular
Cleavage: Perfect in two directions
Fracture: Splintery
Mohs Scale: 4.5–5

Colors: White, tan, gray, blue
Luster: Vitreous, silky
Streak: White
Transparency: Translucent

Description: Pectolite derives its name from the Greek *pektos*, meaning "compacted," a reference to its compact structure. Most pectolite is white, gray, or brown, but there is a sky blue variety called larimar, which is found in the Dominican Republic. Some pectolite can be triboluminescent, giving off light with friction.

North American Locations: Canada: Mountains in Nova Scotia. United States: Mountains in California, Colorado, New Jersey, and Oregon.

Collecting Tips: Pectolite can be found in hydrothermal veins in basalt, andesine, and other rocks associated with zeolite minerals. It is recognized in the field by its compact radiating crystals. Be careful as these fibrous crystals can be very sharp; wear gloves when collecting this mineral. It is also soft, so care must be taken when extracting specimens from host rock.

Processing Tips: Pectolite can take a polish, and carvings and cabochons are often made for collectors. This mineral can splinter easily, so care must be taken when polishing. Many specimens can be left as is, cleaned with soap and water, and put on display.

PERIDOT

Category: Nesosilicate
Crystal System: Orthorhombic
Crystal Habit: Tabular
Cleavage: Fair in two directions
Fracture: Uneven to conchoidal
Mohs Scale: 6.5–7

Colors: Green, yellow
Luster: Vitreous
Streak: White
Transparency: Transparent, translucent

Description: Peridot is the gem variety of the mineral olivine. It has long been used as a gemstone. Peridot is the birthstone for August.

North American Locations: Canada: Mountains in British Columbia. United States: Mountains and deserts in Arizona, Colorado, and New Mexico.

Collecting Tips: Peridot will be easily identified in the field by its bright green color. Many crystals are found damaged by natural causes, but rare, large terminated crystals are occasionally found. Use care when extracting specimens so as not to damage them.

Processing Tips: Rare large crystals can be carved and cut en cabochon, and they also make excellent faceted gems. They can also be tumbled. Specimens with clusters of small crystals make for excellent display pieces and can be trimmed to better show off the peridot.

PETOSKEY STONE

Category: Fossil
Mohs Scale: 3.5–4
Colors: Gray, white

Luster: Vitreous, pearly
Transparency: Opaque

Description: Petoskey stone is a type of fossilized coral. It comes from deposits of Devonian growths of rugose coral, *Hexagonaria percarinata*. In 1965, it was adopted as the state stone of Michigan. It is named after the Ottawa chief Pet-O-Sega.

North American Locations: Along lake shorelines and in nearby river gravels across North America. The main collecting area is the eastern shoreline of Lake Michigan; specimens are also found in Illinois, Indiana, Iowa, and Ohio, and in southern Ontario, Canada.

Collecting Tips: Dry Petoskey stone can look like ordinary limestone. When wet it is easily recognized by its gray coloration and the six-sided pattern made by coral growths.

Processing Tips: Petoskey stone is a very soft material and can be easily carved and shaped into decorative items and even cut into cabochons. Simple hacksaws and sandpaper can be used to shape and polish this fossil.

PETRIFIED WOOD

Category: Fossil
Mohs Scale: 6–7
Colors: Often brown or brownish red but can be any
Luster: Vitreous, waxy

Transparency: Opaque; thin sections are translucent

Description: Petrified wood is a fossil in which the original wood has been replaced, usually by such quartz varieties as chalcedony and jasper, and occasionally by opal. Petrified wood was adopted as the state gem of Washington in 1975.

North American Locations: Plains, deserts, lake shorelines, and river gravels across North America. Excellent collecting sites are located in the deserts of Arizona, Oregon, Utah and Wyoming.

Collecting Tips: Petrified wood is easily recognized in the field as it looks like actual wood that is now a rock-hard mineral. Good pieces show cellular replacement and display the wood rings commonly found inside. You will also notice bark markings and knots on the outside.

Processing Tips: Petrified wood replaced by agate, jasper, or opal takes an excellent polish and has many lapidary applications, including tumbling, carving, polishing, sphere-making, and cutting en cabochon. Large, well-prepared specimens make attractive displays.

PHENAKITE

Category: Nesosilicate
Crystal System: Hexagonal/ trigonal
Crystal Habit: Rhombohedral
Cleavage: Poor in one direction
Fracture: Conchoidal

Mohs Scale: 7.5–8
Colors: Colorless, white
Luster: Vitreous
Streak: Colorless
Transparency: Transparent, translucent

Description: Phenakite is a very rare beryllium mineral. It derives its name from the Greek *phenakos*, meaning "deceiver," a reference to the mineral's resemblance to quartz crystals. It is predominately mined for the jewelry and mineral-collector markets.

North American Locations: Canada: Mountains in Ontario. United States: California, Colorado, Maine, and New Hampshire.

Collecting Tips: Phenakite is found in beryllium-rich pegmatites, granite, and mica schists. It can easily be mistaken for quartz but is harder and shinier than quartz. Care should be taken when extracting fragile specimens from host rock.

Processing Tips: Phenakite has a very high refractive index and makes brilliant faceted stones. It can also be cut en cabochon. Fine single crystals are generally cleaned with soap and water or a light acid, mounted, and put in a display.

PLATINUM

Category: Native element
Crystal System: Cubic
Crystal Habit: Crystals are rare
Cleavage: None
Fracture: Hackly
Mohs Scale: 4

Colors: Steel gray, iron black
Luster: Metallic
Streak: Steel gray
Transparency: Opaque

Description: Platinum is a rare element and, while used by humans for thousands of years, was only recognized as a distinct metal in the 1500s. It derives its name from the Spanish *platina del Pinto*, meaning "little silver from the Pinto River." It is used in fine jewelry and has many industrial applications.

North American Locations: Canada: Mountains and river gravels in British Columbia. United States: Mountains and river gravels in Arizona, California, Georgia, North Carolina, and Wyoming.

Collecting Tips: While often mistaken for silver, platinum has a higher specific gravity. The two minerals also have different streaks (platinum has a steel gray streak while silver has a slivery white streak). Platinum is most commonly found by panning for heavy material in waterways.

Processing Tips: Platinum found by the average rockhound will simply be put on display; large pieces are mounted, smaller pieces displayed in vials. The experienced metalsmith can turn platinum nuggets into jewelry.

PREHNITE

Category: Complex silicate
Crystal System: Orthorhombic
Crystal Habit: Tabular, botryoidal masses
Cleavage: Fair in one direction
Fracture: Uneven

Mohs Scale: 6–6.5
Colors: Green, yellow, tan, white
Luster: Vitreous
Streak: White
Transparency: Translucent, transparent

Description: Prehnite is a basic calcium aluminum silicate. It was named after its discoverer, Hendrik van Prehn, a Dutch colonial administrator.

North American Locations: Canada: Mountains in British Columbia and Quebec. United States: Mountains in Connecticut, Michigan, New Jersey, and Virginia.

Collecting Tips: Prehnite is found in volcanic rocks, often with zeolites and calcite. It can also be found in granite. Prehnite is recognized in the field by its green to yellow hue and botryoidal crystal habit. Care must be taken when removing fragile specimens from host rock. Fine crystals should be wrapped in tissue, newspaper, or Bubble Wrap for transport.

Processing Tips: Large masses of prehnite can be cut, polished, carved, cut en cabochon, and even faceted. Prehnite makes excellent material to be set in jewelry. Crystals and botryoidal masses can be cleaned with soap and water or light acids, prepped or mounted, and put on display.

PYRITE

Category: Sulfide
Crystal System: Cubic
Crystal Habit: Cubic, octahedral, pyritohedral
Cleavage: None
Fracture: Uneven to subconchoidal

Mohs Scale: 6–6.5
Colors: Brass yellow
Luster: Metallic
Streak: Greenish black, brownish black
Transparency: Opaque

Description: Pyrite has long been dubbed "fool's gold," as it has tricked many a greenhorn miner into thinking that they have struck it rich when in fact they have found a common iron sulfide. Pyrite derives its name from the Greek *pyrites lithos*, meaning "stone that strikes fire," a reference to the mineral creating a spark when struck by metal.

North American Locations: All geological settings across North America.

Collecting Tips: Pyrite is a common mineral found in hydrothermal veins, igneous rocks, contact metamorphic rocks, and sedimentary rocks. It is recognized in the field by its brassy gold tone and distinct crystal habits. It is much lighter than gold, and while gold will flatten when struck with a hammer, pyrite will be crushed.

Processing Tips: Most pyrite crystals are simply cleaned with soap and water (or possibly light acids) and put on display. Large masses of pyrite can be cut and polished, carved, and even cut en cabochon (although it's not recommended for most jewelry, as it is very brittle).

PYROMORPHITE

Category: Phosphate
Crystal System: Hexagonal
Crystal Habit: Prismatic
Cleavage: Poor in three directions
Fracture: Uneven to subconchoidal
Mohs Scale: 3.5–4

Colors: Green, yellow, orange, brown
Luster: Resinous
Streak: White
Transparency: Translucent, subtransparent

Description: Pyromorphite is a lead phosphate chloride. It derives its name from the Greek *pyr*, meaning "fire," and *morph*, meaning "form," a reference to the way the mineral takes a crystalline form upon cooling after it has melted into a globule. It is a minor ore of lead but is mostly mined as a crystal specimen.

North American Locations: Canada: Mountains in British Columbia. United States: Mountains and deserts in Arizona, Colorado, Idaho, and Nevada, and in the forests of Pennsylvania.

Collecting Tips: Pyromorphite is found as a secondary mineral in oxidized zones of lead deposits. It is recognizable in the field by its green tone and prismatic crystal habit. Mimetite can often be mistake for pyromorphite; distinguishing the two sometimes requires chemical analysis. Care must be taken with removing specimens from host rocks.

Processing Tips: Pyromorphite is soft and has poor cleavage, so it generally doesn't make for good lapidary material. Crystal specimens are typically left in matrix, cleaned with soap and water or light acids, trimmed and prepped, and put on display.

QUARTZ CRYSTAL

Category: Tectosilicate
Crystal System: Hexagonal
Crystal Habit: Prismatic
Cleavage: None
Fracture: Conchoidal
Mohs Scale: 7

Color: Colorless
Luster: Vitreous
Streak: White
Transparency: Transparent, translucent

Description: Quartz crystal, also known as rock crystal, is the colorless, transparent, macrocrystalline variety of quartz. It is one of the most commonly hunted and found minerals by rockhounds. Quartz, or silicon dioxide, is one of the most abundant minerals in the earth's crust.

North American Locations: All geological settings across North America.

Collecting Tips: Quartz crystals form in nearly all silica-rich igneous, metamorphic, and sedimentary rocks. They are recognized in the field by their colorless, transparent prismatic crystals. Specimens should be wrapped (e.g., in Bubble Wrap) before transport. Hard-rock mining tools will be required to remove most specimens, but loose crystals can be found amid weathered and fragmented igneous rock as float material.

Processing Tips: Quartz crystals are very hard and take an excellent polish. They can be tumbled, carved, cut en cabochon, and faceted into gems. Fine crystals and clusters can be cleaned in light acid and put on display.

QUARTZITE

Rock Type: Regional metamorphic
Structure: Crystalline
Major Minerals: Quartz
Minor Minerals: Mica, kyanite, sillimanite
Fossils: None

Colors: Brown, tan, gray, tan, yellow, reddish
Texture: Medium-grained

Description: Quartzite is metamorphosed sandstone that consists mostly of quartz. It is fine-grained, generally uniform in color and texture, and extremely tough and durable. Its name comes from the German *quarz*.

North American Locations: All geological settings across North America.

Collecting Tips: Quartzite is recognizable in the field by the large grains of quartz that make up the rock. It is most often tan but can be found in almost any color.

Processing Tips: Quartzite can be tumbled, carved, and even cut en cabochon for use in jewelry. Large polished pieces are used for countertops.

RED BERYL

Category: Cylosilicate
Crystal System: Hexagonal
Crystal Habit: Prismatic to tabular; radial, columnar; granular to compact massive
Cleavage: Fair in one direction

Fracture: Uneven to conchoidal
Mohs Scale: 7.5–8
Colors: Light to deep red
Luster: Vitreous to resinous
Streak: White
Transparency: Transparent, translucent

Description: Red beryl was originally named bixbite after Salt Lake City miner and mineral dealer Maynard Bixby. There is also a species of mineral called bixbyite, named after the same man. To clear up confusion, bixbite was renamed red beryl. Red beryl is so rare that it is found in only one place in the world: the Thomas Range of Juab County, Utah.

North American Locations: Canada: No known deposits. United States: The Thomas Range, Juab County, Utah.

Collecting Tips: Red beryl is identified by its bright red coloring and hexagonal crystals. Look for vugs found in white volcanic rhyolite. Hard-rock mining tools will be needed.

Processing Tips: Red beryl can be faceted and makes lovely gemstones, but due to its rarity, many people leave the crystal specimens as they are found. Much of the red beryl found is too small for faceting.

RHODOCHROSITE

Category: Carbonate

Crystal System: Hexagonal/trigonal

Crystal Habit: Rhombohedral

Cleavage: Perfect in three directions

Fracture: Uneven

Mohs Scale: 3.5–4

Colors: Pink, red, dark red, brown

Luster: Vitreous, pearly

Streak: White

Transparency: Translucent, transparent

Description: Rhodochrosite is a magnesium carbonate mineral. Its name comes from the Greek *rhodokros*, meaning "rose colored." Rhodochrosite is the state mineral of Colorado.

North American Locations: Canada: Mountains in British Columbia. United States: Mountains in Colorado, Arizona, Montana, Arkansas, and New Mexico. The world's best specimens come from Colorado.

Collecting Tips: Rhodochrosite generally occurs in hydrothermal veins with a high manganese content and is often found in or near silver deposits as a stalactite. It is recognized in the field by its pink hue. Massive pieces can be solid-colored or banded. Rare crystals are transparent. Much care should be taken extracting the fragile crystals.

Processing Tips: Massive rhodochrosite, while soft, makes for a beautiful lapidary material. It can be carved, face polished, and cut en cabochon. It can be tumbled, but not for as long in the first stages as most harder minerals. Fine transparent crystals can be faceted, but they are too fragile for jewelry.

RHODONITE

Category: Inosilicate
Crystal System: Triclinic
Crystal Habit: Tabular
Cleavage: Good in two directions
Fracture: Splintery to uneven
Mohs Scale: 6

Colors: Pink, gray
Luster: Vitreous
Streak: White
Transparency: Translucent

Description: A manganese inosilicate, rhodonite is also a member of the pyroxene group of minerals. Its name derives from the Greek *rhodos*, meaning "rose," a reference to the mineral's many pink hues, which are caused by manganese. Historically, it has been mined as an ornamental gemstone.

North American Locations: Canada: No known deposits. United States: Mountains in southwestern North America, notably in Arizona, California, Colorado, Nevada, and New Mexico.

Collecting Tips: Rhodonite is found in a wide variety of manganese ore deposits. It can also be found as a metamorphic product of rhodochrosite. Rhodonite is recognized in the field by its various pink hues. When weathered out of its original deposit, rhodonite will oxidize black on the exterior. Chip off the edges to expose the pink interior.

Processing Tips: Rhodonite is an excellent lapidary material and can be used for tumbling, carving, face polishing, slabbing, and cutting into cabochons. It is hard enough to be used for most jewelry applications. Rare translucent rhodonite can be faceted into gemstones.

Rock	**RHYOLITE**

Rock Type: Felsic, volcanic, igneous
Structure: Massive, often showing bedding planes and flow lines
Major Minerals: Quartz, potassium feldspar

Minor Minerals: Amphibole, biotite, glass, plagioclase feldspar
Fossils: None
Colors: Gray, pink, brown, green
Texture: Fine-grained

Description: Rhyolite is a common volcanic rock. Many silicate minerals can be found in vugs and veins in rhyolite, and it is a common rock to find thunder eggs and geodes in.

North American Locations: Mountains and deserts across North America.

Collecting Tips: Rhyolite can be found as massive deposits, breccias, plugs, and dikes. It has the finest grain and lightest color of all volcanic rocks and is often found with obsidian.

Processing Tips: Rhyolite can be tumbled, carved, and cut en cabochon. Banded rhyolite in tan, buff, yellow, and pink colors is cut into decorative objects and countertops.

ROSASITE

Category: Carbonate
Crystal System: Monoclinic
Crystal Habit: Acicular, botryoidal, mammillary, encrustations
Cleavage: Good in two directions
Fracture: Splintery to uneven

Mohs Scale: 4.5
Colors: Blue, green, blue green, colorless
Luster: Vitreous, silky
Streak: Light blue, light green
Transparency: Translucent

Description: Rosasite is a copper zinc carbonate hydroxide and is considered a useful ore of both copper and zinc. It derives its name from the Rosas Mine in Sardinia, Italy, where it was discovered in 1908.

North American Locations: Mountains and deserts in western North America, notably in Arizona, Montana, Nevada, and New Mexico, and in British Columbia, Canada.

Collecting Tips: Rosasite is found exclusively as a secondary mineral in oxidation zones of copper zinc deposits. It can be recognized in the field by its bright blue color. It is often mistaken for turquoise but is quickly distinguished by its botryoidal crystal habit. Care should be taken when extracting fragile specimens from host rock.

Processing Tips: Most rosasite specimens are kept as found. Rosasite can be cut and polished, but it is very soft and not recommended for most jewelry applications. Rough material can be cleaned, trimmed and prepped, and displayed.

Mineral

ROSE QUARTZ

Category: Tectosilicate

Crystal System: Hexagonal/ trigonal

Crystal Habit: Massive

Cleavage: None

Fracture: Conchoidal

Mohs Scale: 7

Colors: Pink, rose

Luster: Vitreous

Streak: White

Transparency: Translucent, transparent

Description: Rose quartz is the pink variety of massive pink-toned crystalline quartz. Crystals of rose quartz are rare and mostly found in Brazil. Its pink color is attributed to traces of titanium. Rose quartz can also have inclusions of rutile, which create an optical phenomenon known as a "star."

North American Locations: Canada: No known deposits. United States: Mountains in California, Colorado, Maine, New Hampshire, and South Dakota.

Collecting Tips: Rose quartz is generally found as masses in granite pegmatites. Rose quartz is recognized in the field by its rosy color. It is often massive and transparent, and has a conchoidal fracture. Search for pieces with the most saturation of pink and the least number of fractures if you plan on using them for lapidary purposes.

Processing Tips: Rose quartz can be used for almost any type of lapidary application. It is hard and takes an excellent polish. It can be carved, tumbled, cut en cabochon, and faceted. Rare star rose quartz can be difficult for beginners to orient properly when cutting to best display the optical phenomenon.

RUBY

Category: Oxide
Crystal System: Hexagonal/trigonal
Crystal Habit: Pyramidal, prismatic barrel
Cleavage: None
Fracture: Uneven

Mohs Scale: 9
Color: Red
Luster: Adamantine, vitreous
Streak: Colorless
Transparency: Transparent, translucent, opaque

Description: Ruby is the blood red variety of the mineral corundum. All other colors are considered sapphire. Before diamond mass marketing, ruby was the traditional gemstone used for engagement rings. Ruby is the birthstone for July.

North American Locations: Exclusively in the mountains in Montana and North Carolina.

Collecting Tips: Ruby can be found in metamorphic dolomite marbles, gneiss and amphibolite, and alluvial deposits. It can be identified in the field by its bright red color and hexagonal crystal system. It is very hard and can only be scratched by diamond or a synthetic material that rates a 10 on the Mohs scale.

Processing Tips: Ruby is very hard and can only be shaped and polished by diamond lapidary tools. Crystals can be prepped and exposed in their host rock to better display the encased hexagonal ruby crystal.

Mineral

RUTILE

Category: Oxide
Crystal System: Tetragonal
Crystal Habit: Prismatic
Cleavage: Good in two directions
Fracture: Uneven
Mohs Scale: 6–6.5

Colors: Reddish brown, red, golden yellow
Luster: Adamantine, submetallic
Streak: Pale brown, yellow
Transparency: Transparent, opaque

Description: Rutile is an important titanium oxide mined for use as an anticorrosive for airplanes, spacecraft, and ships. Rutile is mostly found as a microscopic inclusion in other minerals and is the cause for asterism in minerals such as rose quartz, sapphire, and ruby. It derives its name from the Latin *rutilus*, meaning "reddish."

North American Locations: Mountains in Arkansas, California, Georgia, North Carolina, and Pennsylvania, and in ocean shoreline sands of New Jersey, North Carolina, and Florida. Canadian specimens come from the mountains of British Columbia.

Collecting Tips: Rutile commonly occurs as a minor component of rocks such as granite, gneiss, schist, and pegmatites. Rutile crystals can be identified in the field by their metallic dark red tones and slender prismatic crystal habit. Care should be taken when extracting fragile crystals so as not to damage them. Hard-rock tools will be required.

Processing Tips: Fine rutile crystals are most commonly kept as found, cleaned, prepped, and put on display. Rare transparent crystals can be faceted and used in jewelry.

SANDSTONE

Rock Type: Continental, detrital, sedimentary
Structure: Fine- to coarse-grained
Major Minerals: Quartz, feldspar
Minor Minerals: Calcite, hematite

Fossils: Vertebrates, invertebrates, plants
Colors: Cream, brown, tan, red
Texture: Fine- to medium-grained with angular to rounded grains

Description: Sandstone is abundant across North America. It varies widely in grain size and texture, and often exhibits distinct bedding. Sandstone and limestone are the most common sedimentary rocks.

North American Locations: Mountains, deserts, and plains across North America.

Collecting Tips: Sandstone is collected for its range of colors, textures, and bedding patterns. Sandstone can often contain large fossils, such as the bones of dinosaurs.

Processing Tips: Sandstone requires no special cleaning. Sandstone slabs with intricate multi-colored bedding patterns are used for display purposes. Some fine-grained varieties, especially those with wavy patterns, are cut and polished into decorative objects.

Mineral	SAPPHIRE

Category: Oxide
Crystal System: Hexagonal/trigonal
Crystal Habit: Pyramidal, prismatic barrel
Cleavage: None
Fracture: Uneven to conchoidal

Mohs Scale: 9
Colors: Any except red
Luster: Adamantine, vitreous
Streak: Colorless
Transparency: Transparent, translucent, opaque

Description: While ruby is the bright red variety of the corundum family, sapphire includes all other colors of this gemstone—most commonly blue. Sapphire is the birthstone for September. The name sapphire comes from the Greek *sappheiros*, meaning "blue."

North American Locations: Canada: Mountains in Nunavut. United States: Mountains of Montana and North Carolina, and to a lesser extent in the mountains in California, Colorado, and Massachusetts.

Collecting Tips: Sapphire can be found most commonly in metamorphic rocks, but also in quartz-deficient igneous rocks. It can also be found as an alluvial deposit in rivers and creeks. Sapphire is very heavy, and can be collected using gold pans and fine-mesh screens. Tweezers help pick small crystals out of concentrates.

Processing Tips: Sapphire can be heat-treated in specialty laboratories to better enhance its natural coloration. Smaller crystals are best kept in a small glass container. Larger crystals can be faceted. Faceted gems can be used in jewelry.

SCAPOLITE

Category: Tectosilicate
Crystal System: Tetragonal
Crystal Habit: Square tabular, prismatic, massive
Cleavage: Fair in two directions
Fracture: Uneven to conchoidal
Mohs Scale: 5–6

Colors: Colorless, white, yellow, gray, orange, pink
Luster: Vitreous
Streak: White
Transparency: Translucent, transparent, opaque

Description: Scapolite, once classified as a single mineral, is now a mineral group that includes two very similar minerals, marialite and meionite. Scapolite's name is derived from the Greek *skapos*, meaning "rod," a reference to the mineral's long crystal habit.

North American Locations: Canada: Mountains in British Columbia. United States: Mountains in California and Colorado, and in the hills of Connecticut.

Collecting Tips: Scapolite occurs in metamorphosed limestone, skarn, gneiss, hornfels, granite pegmatites, and sometimes basalt. In North America, specimens tend to be in white to gray tones. It can be identified in the field by its crystal habit, and the fact that it is often fluorescent under both long-wave and shortwave ultraviolet light. Care should be taken when extracting fragile crystals from matrix.

Processing Tips: Colorless crystals can be faceted into gemstones and cabochons for jewelry. Clusters of crystals in matrix make fine display specimens. The matrix may need to be trimmed for a better display.

Mineral # SCHEELITE

Category: Tungstate
Crystal System: Tetragonal
Crystal Habit: Dipyramidal
Cleavage: Fair in one direction
Fracture: Uneven to conchoidal
Mohs Scale: 4.5–5

Colors: Colorless, white, yellow
Luster: Vitreous, adamantine, greasy
Streak: White
Transparency: Translucent, transparent

Description: Scheelite is an ore of tungsten, a crucial metal for industrial purposes. It was named in honor of German-Swedish chemist Carl Wilhelm Scheele, who is credited for discovering tungsten. Scheelite has a very high light refraction and was once used as a diamond imitation but is ultimately too fragile for jewelry.

North American Locations: Canada: Mountains in British Columbia. United States: Mountains and deserts in Arizona, California, Colorado, and New Mexico.

Collecting Tips: Scheelite occurs commonly in metamorphic deposits, hydrothermal veins, and sometimes in granite pegmatites. It can be identified in the field by its crystal habit, heavy weight, and fluorescence under ultraviolet light. Care must be taken when extracting brittle crystals from host rock.

Processing Tips: Fine transparent crystals can be faceted for the collector, but the mineral is too soft and brittle for use in jewelry. This tungsten ore is most often collected as mineral specimens. Crystal clusters in matrix make for fine display pieces.

SHALE

Rock Type: Sedimentary
Structure: Usually bedded, separates easily along laminations
Major Minerals: Quartz, feldspar, mica
Minor Minerals: Pyrite, graphite

Fossils: Leaf and marine
Colors: Gray to brown
Texture: Fine-grained

Description: Shale is a laminated sedimentary rock of lake or marine origin. It is also called mudstone because of its very fine particle size. Shale has a tendency to split along its laminations into thin, flat sections. It is an excellent medium for the preservation of fossils.

North American Locations: Mountains, deserts, and plains across North America.

Collecting Tips: Shale is often found with sandstone and limestone formations. It is readily identifiable by its gray color and tendency to separate readily into flat, leaf-like sections along its many laminations. Always search shale laminations carefully for fossils.

Processing Tips: When processing shale specimens, remember that they are often fragile and separate easily along their laminations. Durable, thin, flat sections of shale are a creative alternative to canvas for artistic painting.

SHATTUCKITE

Category: Inosilicate
Crystal System: Orthorhombic
Crystal Habit: Botryoidal, spherulitic, aggregate, globular, reniform, stalactitic
Cleavage: Perfect in two directions

Fracture: Uneven
Mohs Scale: 3.5
Color: Blue
Luster: Dull, silky
Streak: Blue
Transparency: Opaque, translucent

Description: Shattuckite is a somewhat rare copper silicate. It derives its name from the Shattuck Mine in Arizona, well known for its fine shattuckite specimens. Shattuckite is mined primarily as a gemstone mineral, as its copper content is too low to mine as a copper ore.

North American Locations: Canada: No known deposits. United States: Mountains and deserts in Arizona.

Collecting Tips: Shattuckite can be found in the oxidized zones of copper-rich, hydrothermal replacement deposits. It is identified in the field by its bright to dark blue tones. It can often be confused for other copper minerals such as chrysocolla, azurite, and turquoise, so get familiar with its different crystal habits to distinguish it from these minerals.

Processing Tips: Solid chunks of shattuckite can take a polish. It can be used in jewelry, although it is generally too soft for rings. Specimens can look stunning when mixed with other copper minerals such as azurite, cuprite, and chrysocolla.

SILVER

Category: Native element
Crystal System: Cubic
Crystal Habit: Cubic, octahedral, dodecahedral, wiry, arborescent
Cleavage: None
Fracture: Hackly

Mohs Scale: 2.5–3
Colors: Silver, white
Luster: Metallic
Streak: Silver, white
Transparency: Opaque

Description: Silver is a precious metal long used in industry, medicine, photography, electronics, jewelry, coins, cutlery, and more. It has the highest reflectivity, thermal conductivity, and electrical conductivity of any metal known. It derives its name from the Anglo-Saxon *seolfor*, meaning "silver."

North American Locations: Canada: Mountains in British Columbia. United States: Mountains in Arizona, Colorado, Idaho, and Nevada; the forests of Michigan.

Collecting Tips: Both native silver and silver ore are fairly rare and generally found in areas with magmatic and hydrothermal activity. Native silver nuggets and wire are identified in the field by their silver coloring. Silver tarnishes easily, and specimens can oxidize to a black color. Silver is also heavy, though lighter than platinum.

Processing Tips: Silver nuggets and wire specimens can be cleaned and put on display. Be careful not to put specimens on display near sulfide minerals, as the sulfur will quickly tarnish them. Silver can be melted and used in a wide variety of jewelry applications.

| Rock | **SLATE** |

Rock Type: Regional metamorphic
Structure: Foliated, laminated
Major Minerals: Quartz, feldspar, mica
Minor Minerals: Pyrite, graphite
Fossils: None

Colors: Various shades of gray or brown
Texture: Fine-grained

Description: Slate is a fine-grained, often foliated, metamorphic rock derived from the alteration of shale. It is usually brown, gray, or dark gray in color. Thin, flat slabs of foliated slate were the original chalkboards, and thin, smooth sections of slate once served as whetstones for sharpening knives.

North American Locations: Mountains across North America.

Collecting Tips: Look for slate in exposures of metamorphic rocks. It is easily identified by its gray color, very fine grain, and distinct foliation.

Processing Tips: Slate can be easily cleaned with water and a brush. Smooth sections of slate are used as a creative alternative to canvas for artistic painting.

SMITHSONITE

Category: Carbonate
Crystal System: Trigonal
Crystal Habit: Botryoidal, rhombohedral, scalenohedral
Cleavage: Perfect in three directions
Fracture: Splintery to uneven

Mohs Scale: 4.5
Colors: White, blue, green, pink, yellow, brown
Luster: Vitreous, pearly
Streak: White
Transparency: Translucent

Description: Smithsonite is a zinc carbonate and was once thought to be hemimorphite. It was named in honor of James Smithson, who founded the Smithsonian Institution. Smithsonite is most commonly mined as a gemstone and specimen mineral and is sometimes known as bonamite.

North American Locations: Canada: Mountains in British Columbia. United States: The premier collecting area for smithsonite is Magdalena in the mountains in New Mexico. Specimens are also found in the mountains of Arkansas, California, Pennsylvania, and Utah.

Collecting Tips: Smithsonite is most commonly found as a secondary mineral formed from primary zinc minerals in zinc deposits. It is recognized in the field by its botryoidal crystal habit and heavy weight. Compare suspect specimens to other rocks of the same size nearby. Care should be taken when removing fragile specimens from host rock.

Processing Tips: Smithsonite is mostly collected as a mineral specimen. Pieces can be cleaned, possibly trimmed, and put on display. It can be faceted, but the mineral is a bit too soft for jewelry.

Mineral

SMOKY QUARTZ

Category: Tectosilicate

Crystal System: Hexagonal

Crystal Habit: Prismatic

Cleavage: None

Fracture: Conchoidal

Mohs Scale: 7

Colors: Brown, gray, black

Luster: Vitreous

Streak: White

Transparency: Transparent, translucent, opaque

Description: Smoky quartz is the brown to black variety of crystalline quartz. It gets its coloration from exposure to natural radiation. Some smoky quartz found on the gem market is actually laboratory-irradiated colorless quartz. Smoky quartz was adopted as the state gem of New Hampshire in 1985.

North American Locations: Mountains across North America.

Collecting Tips: Smoky quartz is generally found in quartz veins and granite pegmatites in both igneous and metamorphic rocks. It is commonly found in areas that have been exposed to natural radioactivity. Smoky quartz is recognized in the field by its prismatic crystal habit and brown or black tones. It's a fairly hard crystal, but care should still be taken when extracting specimens. They should be wrapped (e.g., in Bubble Wrap) for transport.

Processing Tips: Intact and pristine smoky quartz crystals make for excellent display specimens. Broken crystals can be tumbled, carved, and cut en cabochon, and are popular for faceting and use in jewelry.

SODALITE

Category: Tectosilicate
Crystal System: Cubic
Crystal Habit: Massive
Cleavage: Poor in one direction
Fracture: Subconchoidal
Mohs Scale: 5.5–6

Colors: Blue, black, orange, green, yellow, violet
Luster: Dull, vitreous, greasy
Streak: White
Transparency: Transparent, translucent, opaque

Description: Sodalite is a blue-toned mineral in the feldspathoid group. There is a variety of sodalite found in the Upper Peninsula of Michigan that looks like gray rock in regular light but glows a bright yellow to orange when exposed to ultraviolet light. Sodalite derives its name from its high sodium content.

North American Locations: Mountains and forests in Canada: Ice River, British Columbia; Bancroft, Ontario; and Mont-Saint-Hilaire, Quebec. United States: Arkansas, Colorado, Maine, and Michigan.

Collecting Tips: Sodalite forms primarily in nepheline syenite pegmatites and is found in crystallized sodium-rich magmas. It is most often a bright to dark blue, typically with black and white coloration throughout. For fluorescent sodalite, search Michigan beaches at night with an ultraviolet flashlight.

Processing Tips: Sodalite takes an excellent polish and can be used for a variety of lapidary purposes, including tumbling, carving, and cutting en cabochon. It can be used in many jewelry applications.

SPHALERITE

Category: Sulfide

Crystal System: Cubic

Crystal Habit: Tetrahedral, cubic, botryoidal, massive, stalactitic, euhedral, granular

Cleavage: Perfect in six directions

Fracture: Conchoidal

Mohs Scale: 3.5–4

Colors: Brown, reddish brown, yellow, green

Luster: Adamantine, resinous, greasy

Streak: Brown, light yellow

Transparency: Opaque, transparent, translucent

Description: Sphalerite is a zinc sulfide mined for its zinc content. It is often found with rare elements such as cadmium, gallium, and iridium. It derives its name from the Greek *sphaleros*, meaning "misleading," a reference to the mineral being difficult to identify due to its many crystal habits.

North American Locations: Canada: Mountains in British Columbia and Ontario. United States: Mountains and hills in Arizona, Colorado, Missouri, New Jersey, New Mexico, and Tennessee.

Collecting Tips: Sphalerite can be found in zinc-rich regions in a wide variety of rocks such as limestone and basalt as hydrothermal veins, mesothermal veins, and hydrothermal replacements. It is generally recognized by its common brown to red tones. Care should be taken when extracting specimens so as not to damage crystals.

Processing Tips: Rare transparent sphalerite crystals can be faceted into beautiful gems, but the mineral is too soft for most jewelry applications. Crystals in vugs can be trimmed for better display.

SPINEL

Category: Oxide
Crystal System: Cubic
Crystal Habit: Octahedral
Cleavage: None
Fracture: Uneven
Mohs Scale: 7.5–8

Colors: Any
Luster: Vitreous
Streak: White
Transparency: Transparent, translucent, opaque

Description: Spinel is a member of the large spinel mineral group. It derives its name from the Latin *spina*, meaning "thorn," which is a reference to the mineral's crystal habit. It is also said that the name was derived from the Greek word for "spark."

North American Locations: Canada: Mountains in Nunavut, Ontario, and Quebec. United States: Mountains in California, Colorado, Idaho, Nevada, and New York.

Collecting Tips: Spinel forms in metamorphic marble, hornfels, gabbro, and alluvial placer deposits. It is generally recognized in the field for its octahedral crystal habit. It can come in a wide variety of colors and tones. It is often fluorescent.

Processing Tips: Spinel is an excellent mineral for many lapidary purposes, including tumbling, carving, cutting en cabochon, and especially faceting. Octahedral crystals in matrix also make for excellent mineral specimens.

SPODUMENE

Category: Inosilicate

Crystal System: Monoclinic

Crystal Habit: Prismatic, massive, tabular, bladed

Cleavage: Perfect in two directions

Fracture: Uneven

Mohs Scale: 6.5–7

Colors: Colorless, white, gray, pink, purple, green, yellow

Luster: Vitreous, pearly

Streak: White

Transparency: Transparent, translucent

Description: Spodumene is a pyroxene mineral with a high lithium content. It is mined as a mineral specimen, as a gemstone, and for its lithium. Pink spodumene is known as kunzite, named after American gemologist George F. Kunz. Green spodumene is called hiddenite, after William E. Hidden, the mineralogist who discovered it.

North American Locations: Canada: Mountains in Manitoba and Ontario. United States: Mountains in California, Colorado, Maine, North Carolina, and South Dakota.

Collecting Tips: Spodumene can be found in lithium-rich granite pegmatites and aplites (fine-grained granite). It is identified in the field by its long, flat, prismatic crystals. Kunzite will be pink to light violet in color, and hiddenite has an almost emerald tone. Care should be taken when extracting fragile crystals.

Processing Tips: Fine, transparent, intact crystals can be cleaned, mounted, and displayed as is. While it is fairly hard, spodumene is a brittle mineral and best approached by experienced lapidaries. Kunzite should have very limited exposure to sun and high heat.

STAUROLITE

Category: Nesosilicate
Crystal System: Monoclinic
Crystal Habit: Prismatic
Cleavage: Poor in one direction
Fracture: Uneven to subconchoidal
Mohs Scale: 7–7.5

Colors: Brown tones
Luster: Subvitreous, resinous
Streak: White, gray
Transparency: Transparent, translucent, opaque

Description: Staurolite is an aluminum iron silicate mined as a collector's mineral. It derives its name from the Greek *stauros*, meaning "cross," a reference to the mineral's common twined crystal habit. Cross-shaped crystals are often called fairy crosses or fairy stones.

North American Locations: Staurolite is an abundant mineral across the mountains and hills of North America, but the twinned X-crystals are less common. Nice specimens with X-crystals are found in Connecticut, Florida, Georgia, New Hampshire, North Carolina, South Carolina, and Vermont, and Manitoba, Nova Scotia, and Ontario, Canada.

Collecting Tips: Staurolite is found in metamorphosed schist and gneiss. It is easily recognized in the field by its brown, twinned crystals, which are often in a cross or X shape. Take care extracting fragile crystals from host rock.

Processing Tips: Staurolite is most commonly collected as a mineral specimen. Host rock can be etched away to better expose the cross-shaped crystals. Pitted crystals in matrix can be ground down to remove pitting, and cabochons can be cut from slabs of staurolite.

Mineral	**STILBITE**

Category: Tectosilicate
Crystal System: Monoclinic
Crystal Habit: Tabular
Cleavage: Perfect in one direction
Fracture: Uneven
Mohs Scale: 3.5–4

Colors: Colorless, pink, peach, white
Luster: Vitreous, pearly
Streak: White
Transparency: Translucent, transparent

Description: Stilbite is a member of the large zeolite family. It derives its name from the Greek *stilbein*, meaning "luster," a reference to the mineral's pearly cleaved surfaces. Stilbite is mined for collectors as well as for use in water purification and chemical filtering.

North American Locations: Canada: Mountains in Nova Scotia and Quebec. United States: Mountains in Colorado, New Jersey, and Oregon.

Collecting Tips: Stilbite occurs in vugs found in basalt, andesine, and also in hydrothermal veins.

It is recognized in the field by its bowtie-shaped, tabular crystals. Care should be taken when extracting fragile crystal specimens in matrix.

Processing Tips: Stilbite is most commonly collected as a mineral specimen in matrix. Dirty crystals can be cleaned with soap and water. The matrix can be trimmed for a more pleasing display. Stilbite is considered too soft and brittle for most lapidary applications.

SUNSTONE

Mineral

Category: Tectosilicate
Crystal System: Triclinic
Crystal Habit: Euhedral, granular
Cleavage: Good in two directions
Fracture: Uneven
Mohs Scale: 6–6.5

Colors: Colorless, yellow, brown, red, orange, peach, green, blue
Luster: Vitreous
Streak: White
Transparency: Translucent, transparent

Description: *Sunstone* is a term used for microcline and a number of other feldspar minerals. It is called sunstone due to its yellow-toned crystals. The most well-known North American sunstone, Oregon sunstone, contains copper and was adopted as the official state gem of Oregon in 1987.

North American Locations: Canada: Mountains and forests in Ontario. United States: Mountains and forests in Arkansas, New Jersey, New York, North Carolina, Oregon, Pennsylvania, and South Carolina.

Collecting Tips: Sunstone is found in both igneous and pegmatitic igneous rocks. Plagioclase sunstone is recognized in the field by its champagne yellow tone. Intact crystals are rare, and the broken shards look like broken pieces of yellow glass. The variety in Oregon can be red, blue, or green and have inclusions of copper, which create an optical phenomenon known as schiller.

Processing Tips: Sunstone, especially Oregon sunstone, takes an excellent polish and can be used in a wide variety of lapidary applications. It tumbles well, makes excellent carvings, and can be faceted into gemstones used for jewelry and collections.

Mineral | # TALC

Category: Phyllosilicate
Crystal System: Monoclinic
Crystal Habit: Granular, fibrous, or massive
Cleavage: Perfect in one direction
Fracture: Uneven

Mohs Scale: 1
Colors: Green to white
Luster: Pearly
Streak: White
Transparency: Translucent to opaque

Description: Talc is the oldest known white pigment and was used in antiquity. Its name comes from the Arabic *talq*, which means "pure," an allusion to the white color of its powdered form. Talc is the softest mineral, having a Mohs ranking of 1.

North American Locations: Mountains across eastern North America, notably in New Hampshire, New York, and Vermont, and in Quebec, Canada.

Collecting Tips: Talc occurs in veins of metamorphosed rock and is identified by its pearly luster and wavy appearance. Rock picks, chisels, and hammers are used to extract specimens, which are fragile and must be handled carefully.

Processing Tips: Talc specimens must be cleaned only with soap, water, and a soft brush. Talc is very soft, so avoid contact with any hard materials. It shouldn't be machined, cut, or ground.

THOMSONITE

Category: Tectosilicate
Crystal System: Orthorhombic
Crystal Habit: Lamellar, radiating aggregates, acicular, bladed
Cleavage: Perfect
Fracture: Uneven to subconchoidal

Mohs Scale: 5–5.5
Colors: Colorless, white, pink, peach, green
Luster: Vitreous, pearly
Streak: White
Transparency: Transparent, translucent

Description: Thomsonite is a somewhat rare mineral of the large zeolite family. It is highly sought after by zeolite collectors. Thomsonite was named in honor of the Scottish mineralogist and chemist Thomas Thomson.

North American Locations: Canada: Mountains in New Brunswick and the Lake Superior region of Ontario. United States: Mountains in California, Idaho, Michigan, New Jersey, New Mexico, and Oregon.

Collecting Tips: Thomsonite occurs in vugs and hydrothermal veins in basalt and andesine. It is identified by its round crystal habit and pink to peach tones. Great care should be taken when extracting fragile crystals in matrix.

Processing Tips: Thomsonite crystals in matrix are generally kept as specimens. They can be cleaned with soap and water and the matrix trimmed for better display. The Lake Superior variety can be tumbled, but don't let it run for too long in the first stages, as it is somewhat soft.

THUNDER EGG

Category: Silicate
Crystal System: Hexagonal
Crystal Habit: Microcrystalline, amorphous
Cleavage: Varies
Fracture: Varies

Mohs Scale: 6–7
Colors: Any
Luster: Vitreous, waxy, pearly
Streak: Typically white
Transparency: Translucent, transparent, opaque

Description: Thunder eggs, also known as lithophysae, are nodules with rhyolite exteriors. They are usually filled with agate or jasper, but sometimes with opal and other minerals. Their name derives from a Native American legend about two warring thunder spirits who took these mineral oddities from thunderbird nests and used them as projectiles.

North American Locations: Mountains and deserts across western North America, notably in Idaho, Arizona, New Mexico, Oregon, and Utah, and British Columbia, Canada.

Collecting Tips: Thunder eggs are found in large deposits of rhyolite high in silica. They are identified in the field by their nodular shape and often bubbly rhyolite exteriors. Suspect material can be cracked open; it is advisable to cut open specimens with a rock saw rather than a hammer.

Processing Tips: Thunder eggs are often cut in half, both faces polished, and displayed as matched pairs. They are sometimes slabbed and the agate, jasper, and/or opal inside used to cut cabochons for jewelry. Broken bits of thunder eggs can be tumble-polished.

TITANITE

Category: Nesosilicate
Crystal System: Monoclinic
Crystal Habit: Wedge-shaped, prismatic
Cleavage: Fair in two directions
Fracture: Conchoidal

Mohs Scale: 5–5.5
Colors: Green, yellow, brown, black, pink, blue
Luster: Vitreous, greasy
Streak: White
Transparency: Translucent, transparent

Description: Titanite is a rare mineral also known as sphene. It is named for its titanium content. Titanite is mined both as a mineral specimen and for its titanium, which is used in industrial applications.

North American Locations: Mountains across Canada and the United States. Fine specimens come from Alaska, Arizona, California, Connecticut, Maine, Montana, and New York, and Ontario and Quebec, Canada.

Collecting Tips: Titanite most commonly occurs in metamorphic rocks such as marble, gneiss, and schist, as well as in certain igneous rocks. It is recognized in the field by its wedge-shaped, prismatic crystals, usually found in bright green to yellow tones. Much care should be used when extracting rare crystals from host rock to avoid damage.

Processing Tips: Titanite is usually kept as a crystal specimen. It is a medium-hard mineral, but too brittle for most jewelry applications. It can be faceted into a gemstone, mostly for collectors. It can be tumbled but is rarely found in sufficient quantity to fill even a small tumbler.

TOPAZ

Category: Nesosilicate
Crystal System: Orthorhombic
Crystal Habit: Prismatic
Cleavage: Perfect in one direction
Fracture: Subconchoidal
Mohs Scale: 8

Colors: Colorless, blue, pink, brown, green
Luster: Vitreous
Streak: Colorless
Transparency: Transparent, translucent

Description: Topaz is an aluminum fluorine silicate mineral that has been long coveted as a gemstone for its brilliance and durability. It derives its name from the ancient Greek island Topazios (though it was never found there). Topaz is the birthstone for November.

North American Locations: Canada: Hills in Manitoba and Nova Scotia. United States: Mountains in California, Maine, Alaska, and especially Colorado, and in deserts in Arizona.

Collecting Tips: Topaz forms in igneous environments such as granite pegmatites and rhyolite deposits, as well as in alluvial deposits. Topaz crystals found in matrix are identified by their prismatic crystals and hardness. Alluvial crystals are generally rounded, translucent to transparent, and heavier than most similar-looking minerals such as quartz.

Processing Tips: Topaz can be tumbled, carved, and cut en cabochon, and makes brilliant faceted gemstones for jewelry. Fine crystals are also collected as mineral specimens. Topaz can be irradiated and/or heated to turn colorless or pale crystals blue.

TOURMALINE

Category: Cyclosilicate
Crystal System: Hexagonal
Crystal Habit: Prismatic, acicular
Cleavage: None
Fracture: Uneven
Mohs Scale: 7–7.5

Colors: Any
Luster: Vitreous
Streak: White
Transparency: Translucent, transparent, opaque

Description: *Tourmaline* is the term used for a large family of hexagonal borosilicate minerals that share a common crystal structure. It can be found in just about every color and tone but is most commonly opaque and black (called schorl). Pink tourmaline is the alternative birthstone for October.

North American Locations: Canada: Mountains in Ontario. United States: Mountains in California, North Carolina, Connecticut, and Maine.

Collecting Tips: Most varieties of tourmaline can be found in granitic pegmatites. It can also be found in metamorphic rocks and rhyolite. Tourmaline is identified by its long, prismatic crystal habit. Care should be taken when extracting rare fragile crystals.

Processing Tips: Tourmaline can be used in a wide variety of lapidary applications. It tumbles well, but carving can be difficult. It is popular for jewelry when cut into slices, cut en cabochon, or faceted.

TURQUOISE

Category: Phosphate
Crystal System: Triclinic
Crystal Habit: Massive, nodular
Cleavage: None
Fracture: Conchoidal
Mohs Scale: 5–6

Colors: Blue, green
Luster: Waxy, subvitreous
Streak: Bluish white
Transparency: Opaque

Description: Turquoise is a hydrated copper phosphate mineral. It derives its name from the French *turquois*, meaning "Turkish," as the mineral was first introduced to Europe by Turkish traders bringing the material from turquoise mines in Persia. It is the birthstone for December.

North American Locations: Canada: No known deposits. United States: Deserts in Arizona, California, Colorado, Nevada, and New Mexico.

Collecting Tips: Turquoise is found as an alteration mineral in hydrothermal replacement deposits, generally in arid, copper-rich regions. It is identified in the field by its bright blue tones, although it can also be found in green tones. Chrysocolla can sometimes be mistaken for turquoise as they are both copper minerals with similar coloration.

Processing Tips: While turquoise has long been used as a lapidary material, much of what is dug up is not suitable for cutting and polishing until stabilized, usually with some sort of resin or epoxy.

ULEXITE

Category: Borate
Crystal System: Triclinic
Crystal Habit: Nodular, acicular
Cleavage: Perfect in one direction
Fracture: Uneven
Mohs Scale: 2.5

Colors: Colorless, white
Luster: Vitreous, silky, satiny
Streak: White
Transparency: Transparent to translucent

Description: Ulexite is a hydrated sodium calcium borate hydroxide. A variety of ulexite known as TV rock forms with masses of thin parallel crystals. When polished at the top and bottom of the crystals, light will transmit through the crystals, causing any image placed at the bottom of the mineral to appear on the top.

North American Locations: Canada: Forests in New Brunswick and Nova Scotia. United States: Deserts and dry lakes in California and Nevada, and plains in Oklahoma.

Collecting Tips: Ulexite is found in dry lake saline deposits and is identified by its parallel fibrous crystals. Care should be taken when collecting, as the crystals can easily break off and become slivers. Wear a face mask when collecting ulexite, as it can be quite dusty.

Processing Tips: Ulexite is too soft for most lapidary applications. It can be easily shaped and polished with sandpaper. Keep ulexite out of water as it can be dissolved.

VANADINITE

Category: Phosphate
Crystal System: Hexagonal
Crystal Habit: Hexagonal prisms
Cleavage: None
Fracture: Uneven
Mohs Scale: 3–4

Colors: Red to reddish brown
Luster: Resinous, adamantine
Streak: Light yellow
Transparency: Translucent

Description: Vanadinite is a hexagonal mineral and a member of the apatite group. It is an important ore of vanadium, for which the mineral is named. Vanadium is mainly used to add toughness to steel.

North American Locations: Canada: No known deposits. United States: Deserts in Arizona and Nevada.

Collecting Tips: Vanadinite occurs as a secondary mineral in oxidization zones of lead ore deposits. It is recognized in the field by its flat, bright red, hexagonal crystals. Extreme care should be taken when extracting the soft, brittle specimens from host rock. Vanadinite has a lead content and should be handled with care.

Processing Tips: Vanadinite is too soft and brittle for most lapidary applications. Crystals and crystal clusters are often kept as mineral specimens. Crystals can darken and lose transparency with prolonged exposure to light, so display them somewhere dark, or keep them in a lightproof container.

VARISCITE

Category: Phosphate
Crystal System: Orthorhombic
Crystal Habit: Cryptocrystalline aggregate
Cleavage: Good in one direction
Fracture: Uneven to splintery

Mohs Scale: 3–4
Colors: Green tones
Luster: Vitreous, waxy
Streak: White
Transparency: Opaque, translucent

Description: Variscite is an aluminum phosphate mineral that is primarily mined for collectors and lapidaries. It derives is name from the former German district Variscia (now called Vogtland), where the mineral was first discovered.

North American Locations:
Canada: No known deposits.
United States: Deserts in Nevada and Utah, and the mountains in Arkansas, North Carolina, Pennsylvania, and Virginia.

Collecting Tips: Variscite occurs as a secondary mineral in hydrothermal replacement deposits and also brecciated sandstones. It is identified in the field by its green tones and crystal habit. It can be found as both nodules and vein material. Variscite is somewhat soft, so care should be taken extracting material from host rock with hard-rock tools.

Processing Tips: Variscite takes a good polish and, while a bit soft, is a popular material in jewelry. It can be carved and cut en cabochon. Nodules and vein material are often cut in half, polished, and displayed.

VESUVIANITE

Category: Sorosilicate
Crystal System: Tetragonal
Crystal Habit: Prismatic, massive, columnar, grainy, acicular
Cleavage: Poor in one direction
Fracture: Uneven to conchoidal

Mohs Scale: 6–7
Colors: Green, yellow
Luster: Vitreous, resinous
Streak: White, pale green
Transparency: Translucent, transparent

Description: Vesuvianite derives its name from Mount Vesuvius, where fine crystals have been found. It is also commonly called idocrase. A blue variety called cyprine, colored by copper, is found in New Jersey. A green variety in northern California is often called californite.

North American Locations: Canada: Mountains in British Columbia and Quebec. United States: Deserts in California and Nevada, and in the mountains in Montana, New Jersey, Maine, and Utah.

Collecting Tips: Vesuvianite occurs most often in metamorphic rocks such as hornfels, limestone, serpentine, and skarns, as well as in igneous environments. It can also sometimes be found in pegmatites. Massive and alluvial vesuvianite can be difficult to identify in the field, as it resembles quartzite, jadeite, and a few other minerals. Rare crystals are recognized by their crystal habits and green to yellow color.

Processing Tips: Massive and alluvial vesuvianite are excellent lapidary materials and can be used for carving and cabochons in jewelry. Rare large crystals can be faceted. Crystals can also be cleaned, possibly prepped and trimmed, and put on display.

WARDITE

Category: Phosphate
Crystal System: Tetragonal
Crystal Habit: Pseudo-octahedral, striated, radial, fibrous
Cleavage: Perfect in one direction
Fracture: Conchoidal

Mohs Scale: 5
Colors: Colorless, white, green, yellow
Luster: Vitreous
Streak: White
Transparency: Translucent, transparent, opaque

Description: Wardite is a hydrous aluminum phosphate hydroxide mostly mined as a mineral specimen for collectors. It is one of the few minerals known in the tetragonal trapezohedral class of minerals. It was named in honor of Henry Augustus Ward, an American professor, mineral collector, and dealer.

North American Locations: Canada: Mountains in Yukon Territory. United States: Deserts in California, Nevada, and Utah, and in the mountains in Maine, and New Hampshire.

Collecting Tips: Wardite occurs in low-temperature phosphatic nodules in sedimentary rock deposits. It is recognized in the field by its color and rare pseudo-octahedral crystal habit. Great care should be taken when extracting fragile specimens. Wrap them (e.g., in Bubble Wrap) for transport.

Processing Tips: Wardite is collected as a rare specimen. The matrix can be trimmed for a more pleasing display. Rare colorless crystals can be faceted for collectors, but wardite is too soft for most jewelry applications.

WAVELLITE

Category: Phosphate
Crystal System: Orthorhombic
Crystal Habit: Spherical, radial aggregates
Cleavage: Perfect in one direction
Fracture: Uneven to subconchoidal

Mohs Scale: 3.5–4
Colors: Green, yellow, brown
Luster: Vitreous, resinous, pearly
Streak: White
Transparency: Translucent

Description: Wavellite is an aluminum phosphate primarily mined as a mineral specimen and gemstone. It was named in honor of William Wavell, an English physician, botanist, historian, and naturalist.

North American Locations:
Canada: No known deposits. United States: Forests in Arkansas, Alabama, Virginia, and Tennessee, and in the mountains in Colorado and Nevada.

Collecting Tips: Wavellite occurs as a secondary mineral in oxidized zones of low-grade metamorphic rocks, phosphate-rich sedimentary rocks, and epithermal veins. It is identified in the field by its bright green color and spherical crystal habit. It is a somewhat soft mineral and also brittle, so care must be taken when extracting specimens from host rock.

Processing Tips: Wavellite is most commonly collected as a mineral specimen. It can take a polish and be used in some jewelry applications, but is soft and brittle, so not recommended for use in rings. Heavily stained specimens can be cleaned in a light acid bath.

WULFENITE

Category: Molybdate
Crystal System: Triclinic
Crystal Habit: Tabular, prismatic
Cleavage: Good in one direction
Fracture: Subconchoidal
Mohs Scale: 2.5–3

Colors: Orange, yellow, red
Luster: Adamantine, resinous
Streak: White
Transparency: Transparent, translucent

Description: Wulfenite is a lead molybdenum mineral mined for specimens. Though it is a minor ore of lead and molybdenum, there are more economical resources available. It was named in honor of Austrian mineralogist Franz Xaver Freiherr von Wulfen.

North American Locations: Canada: Mountains in Quebec. United States: Deserts in Arizona, California, and Nevada, and in the mountains of Colorado, Idaho, and Montana.

Collecting Tips: Wulfenite occurs in hydrothermal replacement lead deposits as a secondary mineral. It is recognized in the field by its orange to yellow flat, tabular to prismatic crystals. Some specimens will glow dark yellow to orange under shortwave ultraviolet light and orange to red under long-wave ultraviolet light. It is soft and fragile, so use care when extracting crystals.

Processing Tips: Wulfenite crystals are soft and brittle, and often don't form large enough to facet. Rare large crystals can be faceted for collectors, but are not recommended for use in jewelry.

ZIRCON

Category: Nesosilicate
Crystal System: Tetragonal
Crystal Habit: Prismatic, dipyramidal, tabular, massive
Cleavage: Fair in two directions
Fracture: Uneven to conchoidal

Mohs Scale: 7.5
Colors: Any
Luster: Vitreous, adamantine, greasy
Streak: White
Transparency: Transparent, translucent, opaque

Description: Zircon is a zirconium orthosilicate and has been collected and mined as a gemstone for thousands of years. It has a brilliance and luster comparable to diamonds. The name is thought to be derived from either the Arabic *zarkun*, meaning "cinnabar" or "vermillion," or the Persian *azargun*, meaning "gold colored."

North American Locations: All geological settings across North America. Large crystals are found in the mountains of California, Colorado, Maine, and New Jersey, and in the hills of Oklahoma. Canadian specimens come from the hills and mountains of British Columbia, Nova Scotia, Ontario, and Quebec.

Collecting Tips: Zircon occurs as an accessory mineral in most granitic rocks, schist, and metamorphosed igneous rocks. It is identified in the field by its crystal habits. It is relatively heavy and can sometimes be panned like gold from placer deposits.

Processing Tips: Zircon has a high light refraction, and faceted gems rival diamonds in brilliance. If collected in large enough quantities, it can be tumble-polished. Fine crystals can be displayed.

ZOISITE

Category: Sorosilicate
Crystal System: Orthorhombic
Crystal Habit: Prismatic
Cleavage: Good in one direction
Fracture: Uneven, conchoidal
Mohs Scale: 6–7

Colors: White, gray, yellow, pink, blue
Luster: Vitreous, pearly
Streak: White, colorless
Transparency: Transparent, translucent

Description: Zoisite is a family of calcium aluminum hydroxy sorosilicates, most of which are mined as gemstones. This family includes tanzanite (blue) and thulite (pink). It was once called saualpite but was renamed in honor of Austrian scholar Sigmund Zois.

North American Locations: Canada: Mountains in British Columbia, Ontario, and Quebec. United States: Deserts in Arizona and California, and the mountains in Alaska, North Carolina, and Virginia.

Collecting Tips: Zoisite occurs in such regionally metamorphosed rocks as eclogites and blueschist facies. The most common type of zoisite in North America is thulite. It can form small crystals and also be massive. Hard-rock tools will be required for removing massive thulite from host rock. Take care when extracting the fragile crystals.

Processing Tips: Massive pink thulite takes an excellent polish and can be used for many lapidary projects such as tumbling, carving, and cutting en cabochon. It is hard enough for use in a variety of jewelry applications. Crystals in matrix can be cleaned, possibly trimmed, and displayed.

Appendix

ADDITIONAL RESOURCES

Websites

https://geology.com
https://geologyscience.com
www.mindat.org
www.minerals.net
www.webmineral.com

Books

Garlick, Sarah. *National Geographic Pocket Guide to Rocks and Minerals of North America*. National Geographic, 2014.

Pellant, Chris. *Rocks and Minerals*. Dorling Kindersley, 2002.

Pough, Frederick H. *Peterson Field Guides: A Field Guide to Rocks and Minerals*. 5th ed. Houghton Mifflin Harcourt, 1998.

Prinz, Martin, George Harlow, and Joseph Peters, eds. *Simon & Schuster's Guide to Rocks and Minerals*, 1978.

Rock Shops

Crystal Barista, Salt Lake City, Utah
Dave's Down to Earth Rock Shop, Evanston, Illinois
Nevada Mineral & Book Company, Orange, California
Real Earth Creations, Mount Ida, Arkansas
The Rock Hut, Leadville, Colorado

Rockhounding Shows/Events

Denver Gem & Mineral Show, Denver, Colorado (September)
Houston Gem & Mineral Society Annual Show, Houston, Texas (November)

New England Gem & Mineral Show, Topsfield, Massachusetts (June)
NJ Mineral, Fossil, Gem & Jewelry Show, Edison, New Jersey (August)
Powwow Gem & Mineral Show, Quartzsite, Arizona (January)
Rockhound Gemboree, Bancroft, Ontario, Canada (August)
Tucson Gem and Mineral Show, Tucson, Arizona (February)

Rockhounding Organizations

American Federation of Mineralogical Societies
www.amfed.org

California Federation of Mineralogical Societies
https://cfmsinc.org

Eastern Federation of Mineralogical and Lapidary Societies, Inc.
https://efmls.org/

Midwest Federation of Mineralogical and Geological Societies
www.amfed.org/mwf

Northwest Federation of Mineralogical Societies
www.northwestfederation.org

Rocky Mountain Federation of Mineralogical Societies
www.rmfms.org

Southeast Federation of Mineralogical Societies
www.amfed.org/sfms/

GLOSSARY

Aventurescence
A metallic glitter effect seen in some gems and minerals.

Axis
An imaginary line between the opposing sides of a crystal that is used to identify its system and define its shape.

Cabochon
A gem or mineral cut in convex form and polished but not faceted.

Chatoyancy
An optical effect in which certain polished gems reflect a single thin ray of bright light.

Cleavage
The way a gem or mineral breaks along certain planes according to its crystal structure.

Conchoidal
Shell-shaped; used to describe a fracture pattern.

Crystal
The particular shape in which many minerals form.

Crystal habit
The external shape of a gem/mineral crystal or group of crystals.

Crystal system
A seven-category classification of gems/minerals based on their crystal structure. Each crystal system is defined by number and relative length of its crystal axes, and the angles at which these axes intersect.

En cabochon
A gem form that is convex in shape and not faceted.

Facet
An artificially produced, smooth, flat surface on a gem; also, to cut facets on a gemstone.

Felsic
A light-colored igneous rock composed mainly of feldspar minerals.

Float
Material that has moved from its point of origin.

Fluorescence
The emission of brightly colored light by a rock, gem, or mineral when exposed to ultraviolet light.

Fossil

Remains or impressions of ancient organisms preserved in a rock or other natural material.

Fracture

The way a gemstone or mineral breaks other than along cleavage planes. Common fractures include conchoidal, subconchoidal, splintery, and uneven.

Gem

A mineral or mineral-like material that has been fashioned into a form suitable for jewelry use.

Host rock

The rock surrounding a gem, mineral, or fossil.

Igneous

Formed by the direct solidification or crystallization of magma; one of the three primary classifications of rock.

Inclusion

A foreign body (e.g., another mineral or bubble) within a crystal.

In the rough

In a natural state; without any treatment, such as polishing.

Iridescence

A play of vivid colors on the surface of a gem or mineral.

Knapping

The process of chipping away rock to create points, knives, etc.

Labradorescence

A play of color caused by selective internal reflections.

Lapidary

Can refer to the act of cutting, polishing, and engraving rocks, gems, minerals/mineraloids, and fossils, as well as to the person doing lapidary work.

Luster

A reference to the shine of a gem or mineral. Common lusters include vitreous (shines like broken glass), pearly (shines like a pearl), and greasy (shines as though coated in oil).

Mafic
A dark-colored igneous rock rich in iron and magnesium minerals.

Magma
Molten rock originating beneath the earth's crust that solidifies to form igneous rocks.

Matrix
A mass of rock in which gems, minerals, or fossils are found. During a lapidary process, a gem, mineral, or fossil can be cut "in matrix," so that it remains in the mass of rock but is more visible.

Metamorphic
Formed through the alteration of pre-existing rock, usually by heat and pressure; one of the three primary classifications of rock.

Mineral
A natural, solid, inorganic, crystalline material with a definite chemical composition.

Mineraloid
A mineral-like material that lacks a crystal structure or definite composition, such as opal and obsidian.

Mohs scale
A 1–10 scale of gem or mineral hardness, 1 being the softest (talc) and 10 being the hardest (diamond).

Nepheline-syenite
An igneous rock consisting mostly of feldspar minerals.

Ore
A mineral containing a valuable constituent for which it is mined.

Outcrop
The section of a subterranean rock formation that is exposed on the surface.

Pay dirt
Earth or ore containing valuable material.

Pegmatite
A body of very coarse-grained igneous rock, often containing rare minerals and well-developed crystals.

Placer
A deposit of valuable heavy minerals such as gold that has become concentrated in loose sediments.

Preform
A roughly shaped cabochon.

Pseudomorph
A mineral having the characteristic outward form of another species.

Rock
A mass of mineral material. There are three main categories of rock: igneous, sedimentary, and metamorphic.

Schiller phenomenon/effect
The brilliant play or sheen of colors within a crystal.

Sedimentary
Formed by the accumulation of sediments; one of the three primary classifications of rock.

Star asterism
Also simply "star," or "asterism," a star asterism is a star-shaped figure exhibited by some crystals when light reflects from inclusions.

Streak
The color of a gem or mineral in powder form. A "streak test" is accomplished by rubbing the gem/mineral against an unpolished piece of porcelain.

Tailing
Discarded or refuse material produced by the mining or processing of mineral ores.

Transparency
A reference to the amount of light that can pass through a gem or mineral. Transparency can range from transparent (passes all light) and translucent (passes some light) to opaque (passes no light).

Tumble
To rotate rocks, gems, or minerals/mineraloids in a tumbling barrel until polished.

Vein
A thin fracture in a rock that is filled with secondary minerals.

Vug
A cavity in a rock that is often filled with crystals.

INDEX

About the Authors

Lars W. Johnson was born and raised in the Northwest, and has been an avid rockhound since his childhood. He is dedicated to inspiring enthusiasm and inclusivity to those new to rockhounding, and he has a curiosity for locating, collecting, and sharing experiences that seasoned rockhounds can appreciate.

Stephen M. Voynick is a Colorado-based mineral collector and former hard-rock miner. He is the author of ten books and hundreds of magazine articles, most on topics of minerals, mineral collecting, mining, and mining history.